SEX

in the

SEA

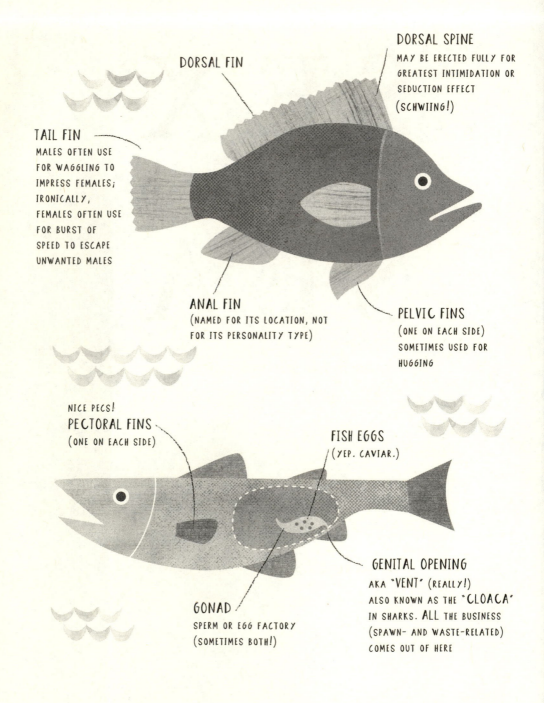

DORSAL FIN

DORSAL SPINE
MAY BE ERECTED FULLY FOR GREATEST INTIMIDATION OR SEDUCTION EFFECT (SCHWIING!)

TAIL FIN
MALES OFTEN USE FOR WAGGLING TO IMPRESS FEMALES; IRONICALLY, FEMALES OFTEN USE FOR BURST OF SPEED TO ESCAPE UNWANTED MALES

ANAL FIN
(NAMED FOR ITS LOCATION, NOT FOR ITS PERSONALITY TYPE)

PELVIC FINS
(ONE ON EACH SIDE) SOMETIMES USED FOR HUGGING

NICE PECS!
PECTORAL FINS
(ONE ON EACH SIDE)

FISH EGGS
(YEP. CAVIAR.)

GONAD
SPERM OR EGG FACTORY (SOMETIMES BOTH!)

GENITAL OPENING
AKA "VENT" (REALLY!) ALSO KNOWN AS THE "CLOACA" IN SHARKS. ALL THE BUSINESS (SPAWN- AND WASTE-RELATED) COMES OUT OF HERE

SEX

in the

SEA

Our Intimate Connection with
Sex-Changing Fish, Romantic Lobsters,
Kinky Squid, and Other Salty
Erotica of the Deep

MARAH J. HARDT

Illustrated by Missy Chimovitz

St. Martin's Press
New York

www.stmartins.com

Portions of the text in Chapter 5 first appeared in "Getting to Know Whale Vaginas in 7 Steps," *Scientific American* (blog), June 11, 2014.

Parts of the text in the Chapter 5 section "Dwarf and Dominatrix" first appeared in "All Female Bone-Devouring Worms Fancy Dwarf Males, Except One," *Deep Sea News,* December 14, 2014.

Portions of the text in Chapter 7 first appeared in "Hybrid Corals: Sex Gone Awry or Saving Grace?" *Scientific American* (blog), September 25, 2014.

Design by Letra Libre, Inc.

Library of Congress Cataloging-in-Publication Data

Hardt, Marah J., 1978–
    Sex in the sea : our intimate connection with sex-changing fish, romantic lobsters, kinky squid, and other salty erotica of the deep / Marah J. Hardt.
        pages   cm
    Includes bibliographical references and index.
    ISBN 978-1-137-27997-2 (hardcover)
    ISBN 978-1-4668-7922-5 (e-book)
    1. Marine animals—Sexual behavior. 2. Marine animals—Reproduction.
3. Deep-sea animals—Sexual behavior. 4. Deep-sea animals—Reproduction.
5. Courtship in animals. 6. Sexual behavior in animals. I. Title.
    QL122.H276   2016
    591.77—dc23

                                                                    2015032689

Our books may be purchased in bulk for promotional, educational, or business use. Please contact your local bookseller or the Macmillan Corporate and Premium Sales Department at (800) 221–7945, extension 5442, or by e-mail at MacmillanSpecialMarkets@macmillan.com.

First Edition: February 2016

10   9   8   7   6   5   4   3   2   1

For Steve, who believes.

Siempre.

CONTENTS

Introduction   Getting Your Fins Wet   1

ACT I
DATING GAMES   5

1   The Quest: Seeking Sex in Saltwater   7

2   Luring a Lover: The Art of Salty Seduction   35

3   Flex Your Sex: Sex Change in the Sea   69

ACT II
SEALING THE DEAL   97
Part 1: Sexual Intercourse

4   The Penis Chapter: Sex as a Contact Sport   99

5   Inner Chambers: Influencing Sex from the
    Inside Out   133

ACT II
SEALING THE DEAL   163
Part 2: Sexual Outercourse

6   Oceanic Orgies: Getting It On in Groups   165

7   Synchronized Sex: A Neighborly Affair   183

ACT III
POST-CLIMAX

205

8   Turning Up the Sex Drive: How to Spark
Successful Sex in the Sea   207

Conclusion   The Sea Is One Sexy Beast   233

Acknowledgments   235
Notes   239
Bibliography   259
Resources: Diving A Little Deeper   269
Index   273

GETTING YOUR FINS WET

IT SMELLS LIKE SEX. I COULDN'T SHAKE THAT SOME-what disturbing thought as I floated at the surface, watching the moonlight glisten off the ever-widening slick—the residue of the night's intimacy. The distinct, musty odor was undeniable. I guess I shouldn't have been so surprised. I had just spent the past two hours watching corals spawn. But that event looked so unlike sex that I never expected the aftermath to smell like it.

As I pulled strands of mucous-y coral goo out of my hair, I caught the eye of the other researchers similarly engaged in de-spawning themselves. We smiled at one another with a knowing look. There is a certain level of bonding that occurs when floating amid the leftovers of one of Nature's biggest orgies.

That's the thing about sex in the sea. It is at once utterly foreign, yet there are hints of the familiar—but only *just.* For the most part, sex beneath the waves looks nothing like what we think of as intercourse. That's what happens after several hundred million years of intense battles over who can reproduce the most—evolution gets a little funky.

From the highest reef crests to the deepest trench in the sea, getting laid and not getting eaten are the two biggest concerns for nearly every animal on the planet. Thus Nature invests heavily in the art of both sex and

survival. Life's great purpose—to successfully pass along to future generations one's good looks and all the genes that go with it—relies on both
skills. But not equally. A deft survivor that lives a long but celibate life
loses the evolutionary game; a great lover, adept at attracting and securing
a mate, needs only to survive long enough to get the deed done.

In the end, it all comes down to sex.

Thus the mind-boggling array of ways to seal the deal in Nature. And
I'm not talking *Kama Sutra*–style creative contortionism here. That's just a
bunch of minor postural adjustments. Real sexual innovation occurs in the
wilderness, and nowhere are things more wet and wild than in the ocean.
After all, that's where sex was invented. That's where Mother Nature has
been practicing her procreative creativity the longest.

Under the sea, the missionary position is in the minority. Even the
images of mating animals on land most familiar to us—the neighbor's dog
aggressively humping your leg—are outliers. Instead, sex may look like a
handshake; an interlocking ring of multiple individuals forming a closed
loop of lovers; or it may be a microscopic male squeezing out sperm while
living his days inside his giantess mate's kidney. Peek below the shimmering surface of the sea and witness worm penis jousting matches, full-moon
sex parties, sunset spawning blitzes, and likely the biggest threesome in the
world (with the lovers holding their breath the entire time).

Each of these strategies suits the location and the lifestyle of the species: cold, dark, deep-sea love dens versus warm, bright, reefy love nests;
the social, vocal pods of whales contrasted with the independent, silent
life of sharks; the microscopic quest and size of copepods compared with
the epic journeys of giant bluefin tuna; the daily spawning rush of wrasse
alongside the once-a-lifetime sex fest of salmon. Every version has been
honed to maximize the chance of successful reproduction. It's a salty symphony of sex that ensures the big blue sea stays bountiful year after year,
age after age, right up to the present day. Almost.

Over the last century or more, those myriad instruments of sex have
started to, well, go a bit off-key. This is bad news not just for horny halibut,
but for us as well.

The way marine life gets busy in the deep matters—it matters for food security, human health, coastal development, climate change, and other global issues. Take food security. Nearly three billion people rely on fish as a major source of protein. Half of that fish comes from the sea. To feed that many people requires a lot of fish successfully making a lot of baby fish every year.

But that's not the only reason the sex lives of snapper and sardines are of interest.

Beachgoers and coastal homeowners take note: like the great walls that defended medieval cities, the mighty underwater reefs built by millions of oysters or corals protect and stabilize the shores. These natural barriers break up wave energy, helping to guard the coastline from storms and heavy surges. And because they are living walls, they can grow—rather than erode—over time, rising even as sea levels do. Lose the reef builders and it is not long before the sand starts slipping back into the sea.

It's energetically expensive for animals to build and maintain those giant blocks of reef. Add to that the energy they now have to spend fighting off ocean acidification, pollution, and invasive species, and there is just not much left to invest in sex. It's like the exhaustion that follows long days at the office—sex tends to take a backseat to sleep.

We rely on the rapid reproduction of tiny crustaceans to feed the fish that feed us. We depend on the mass spawning of corals to create the reefs that provide habitat for the millions of species that we use for medicine, food, and simple enjoyment. We rely on the prolific procreation of oysters, clams, and other shellfish to filter and clean our coastal waters. Whether it is finding the next cancer-fighting compound, feeding a growing population, or fueling economies, we depend upon the extraordinary abundance of ocean life to sustain us—and that abundance depends upon lots and lots of sex. Without successful sex in the sea, we're sunk. That's why knowing what actually goes on "down there" matters to us up here.

Studying the sex lives of animals that live far from shore, in places we often cannot go, isn't easy (or cheap). But scientists are getting better at it every day. Just a few years ago we had no idea how many marine species got busy; there were several we didn't even know existed! Today, we are privy to some of their most intimate secrets. Researchers have at their fingertips a far greater tool kit for examining the reproductive acts of ocean inhabitants. These new approaches and technologies have led to an explosion of science's own literature of salty erotica. From the acrobatic sex positions of deep-sea squid to uncovering how many males (at minimum) a female shark *really* sleeps with, scientific papers are now brimming with saucy tales.

This book sets out to explore the best of them, bringing this most recent eruption of research out of the lab and into your lap. The stories cover the age-old progression from landing a first date to consummation and concludes with ideas about how to boost successful sex in the sea. It's a bold ride through the red-light districts of the big blue sea that may leave you feeling slightly humbled, but hopefully inspired. After all, sex is the way life replenishes—it is what fuels growth and abundance and, over time, diversity—nature's insurance policy against the vagaries of the environment. The more we understand sex, the more we can help shepherd this force toward restoring productivity in the sea rather than proving a giant block against it.

Should you walk away with a few ideas on how to enhance your own sexual encounters of the terrestrial kind, consider it a bonus . . .

The events portrayed in the short vignettes that accompany each chapter are works of fiction. The characters within neither represent nor reflect any specific person or individual animal. Imagination and the art of exaggeration have been employed, and literal readers are cautioned to take these tales with a grain of salt. A suggested "sex-sea soundtrack" is provided to help set the mood. Do not try any of these acts at home without the support of the appropriate whale, shark, squid, lobster, fish, or other trained marine professional.

— ACT I —

DATING GAMES

THE SEX LIVES OF MARINE LIFE, JUST LIKE OURS, involve more than just sex. Before two (or three—or a hundred) individuals can get down to business, they first have to get in the mood. Before the sex comes the seduction, but before the seduction comes the search.

The following three chapters take a look at all that goes on to set the stage for strangers to engage in the most intimate act imaginable: the entwining of their DNA. Under the sea, the buildup can be taxing, requiring months of travel and perhaps weeks of preparation. Other times, there's little more than a wink and a nod and a pair of lovers are off and swimming. From long-distance affairs to underhanded ambush attacks, the strategies for finding and securing a mate in the sea are as diverse as the sex itself. We'll start, however, where many animals must start: that daunting task of seeking out a mate in saltwater. The ocean, after all, is a pretty big place to have to find a date.

1

THE QUEST

Seeking Sex in Saltwater

SEX-SEA TRIVIA

- *There are microscopic singles bars in the sea.*
- *Groupers like sex on the edge.*
- *The largest organ on the planet shaped (in part) by female sexual desire isn't a penis.*
- *Male blue whales are getting their Barry White on.*

SEX-SEA SOUNDTRACK

1. "Somebody to Love"—Freddie Mercury
2. "Ain't No Mountain High Enough"—Marvin Gaye
3. "Can't Get Enough of Your Love, Babe"—Barry White
4. "Good Vibrations"—The Beach Boys

AH, THE QUEST FOR A MATE. IT'S NEVER BEEN EASY. But if you think it's hard to find a partner on land, just imagine having to search the entire sea for a suitable lover. Up on terra firma, we've only two dimensions to deal with (people don't fly through the air, after all) and a quarter of the earth's surface (much of which is uninhabited). Plus, we've got the Internet, with dating sites galore all claiming the most sophisticated algorithms to ensure that compatible individuals connect.

The sea represents a far more challenging environment, and not just because of the lack of Match.com. The oceans are just so vast, covering about 99 percent of the habitable space on the planet. Sure, there are some species that live on a small reef, mating with the equivalent of their high school sweetheart. But for many, from whales to wahoo, individuals from a single species are dispersed over entire ocean basins. In this enormous space, even the biggest animal ever to live on Earth—the blue whale—is but a small fish in a very, very big pond. From high above, the sleek silhouette of this two-hundred-ton leviathan disappears into an indiscernible speck long before the edges of the continents come into view. And that's just the superficial perspective. Below the whale lie miles and miles of ocean, through which any number of mates may be swimming—or not. For the smallest members of the ocean—the microscopic grazers and hunters upon which the rest of the food web depends—a bucket of water may as well be the Pacific.

How in the world, within all the vastness that is the sea, do individuals ever find each other? This search for a mate is akin to the quest for the Holy Grail, and like King Arthur, most ocean animals—from the infinitesimal to the enormous—must undertake this mission at least once in their lifetime. And over the eons, enough of them have succeeded.

The swirling spheres of millions of sardines, the mile-long migrating schools of mobula rays, the winding walls of oyster reefs so tall they break the surface at high tide—such astounding abundance is testament to the fact that despite the long odds, species from shrimp to sperm whales continue to find suitable mates. For some, the winning search strategy is to journey to one of the ocean's many hot spots for daily or seasonal soirees. For others, hookups are isolated events, requiring some sophisticated, long-distance advertising in order for one partner to find the other. And in some cases, it's a carefully calculated combination of both: a matter of getting to the right place at the right time, and letting others know you've arrived.

SEARCH TACTIC NO. 1:
HEAD FOR THE SINGLES BARS

He gazed out across the crowded city. She was out there, somewhere, keeping a low profile so as not to draw too much attention to herself. That was smart. But it made his search all the more difficult. He had almost nothing to go on. She had left a string of clues he could follow, if he was careful. This was indeed the most important case of his life, and the most challenging. By now others could also be hot on her trail. His mission was to find her before anyone else did. It would not be easy. Asking too many questions, broadcasting his presence too widely, would only bring unwanted attention. No, this case required the utmost discretion and superior tracking skills. If he missed a single clue, he could be thrown off her trail for weeks.

He knew he didn't have long. They could both be dead within days.

SO MIGHT GO an opening from a Bogart film. But it is equally fitting as a memoir of the dating life of the ocean's smallest residents. Searching for a microscopic mate in the sea is like looking for a needle in a haystack, only the haystack is the size of Mount Everest. Consider copepods, distant relatives of shrimp that fuel the ocean's food web. Some species are as small as a sesame seed; others about as long as your thumbnail. Even within the confines of the average home aquarium, a male copepod swimming around randomly is likely to bump into a female copepod about once per year, yet individuals may live only a few months and some only a few weeks.

With an entire ocean to contend with, how in the world does an animal smaller than a grain of rice find an equally tiny (and transparent) mate in all that blue? Dr. Peter Franks, a professor at Scripps Institution of Oceanography, has a simple answer: they go to singles bars, of course.

Before we dive into the details of miniature singles scenes in the sea, here's why the sex life of copepods is of interest. Small in size, copepods are big in impact. Tiny and packed with fat, they are the baby food of the sea, feeding countless larval crabs, fish, and squid. Crunchy on the outside, full of gooey oils on the inside, they're also the go-to meal for enormous, swirling schools of bait fish—sardines, anchovies, herring—that in turn feed the tuna, snapper, and cod we like to eat. Many of the eleven thousand or so species of copepod are small enough to fit on the tip of a pencil, yet their abundance satiates pods of whales. Attaining those kinds of numbers requires copious amounts of copepod copulation, which requires some close physical contact between males and females.

To close the gap between them, a male must have excellent detective skills, homing in on subtle clues left behind by a swimming female. Dr. Jeannette Yen of Georgia Tech is an expert in the way minute animals move through water, and she explains that when you are as small as a copepod, water behaves differently; it's thicker and stickier. As copepods swim, they must dig their way along, pushing the water out in front of them and leaving temporary tunnels of disturbed water flow behind. Like barely-there footprints in the sand, these are the tracks of the females the males then trace by feeling their way along.

Fine, feathery hairs on copepods can detect subtle differences in water flow. A surge of motion from one direction may indicate a predator; the rippling waves of a swimming female create a different pattern. In some species, females may also infuse their personal corridor with pheromones, making their signal even stronger. Whether by feel or scent, when a male crosses a female's track, he literally flips.

He rapidly spins his body, cartwheeling into the middle of the trail as he begins a frenetic high-frequency zigzag—in three dimensions—across the trail. Once he's locked on, the male's pirouetting pursuit is remarkably tight, and he successfully narrows the distance from up to one hundred body lengths away. That's the equivalent of a guy standing on top of a sixty-story building and picking out his girl down on street level by the smell of her perfume.

These trails last but a few seconds, however, which is where the importance of singles bars—spots of particularly still water in an otherwise swirling sea—come in. Copepods congregate where the footprints of females can last a little longer. That is, in the quiet, thin section of water where two different pieces of ocean meet.

Far from a uniform pool of blue, the ocean is much more like a layer cake. Different water strata of varying temperature or salinity stack up on top of one another throughout the water column, and where two water masses meet, a distinct boundary layer appears. Boundaries such as thermoclines are created by differences in temperature, but such an interface can also be created by differences in salinity or by currents and gyres when swift, spiraling eddies move through one piece of the water column but not another. These carry sections of the sea along at different rates, similar to high-and low-altitude clouds riding different winds.

For copepods swimming through the open sea, the thin boundaries between warm and cold, saltier and fresher, faster and slower water provide distinct "landmarks" within an otherwise featureless blue. These boundaries also remain relatively stable, with little mixing of water across two layers. This means the water *within* the boundary layer—a thin bar—remains relatively still. It is here that the females can pour their eau de copepod into their freshly carved tunnels. Rather than love letters, Yen remarks, copepod females leave "love envelopes." The quieter the water, the longer the message remains.

We may only detect the most extreme contrasts between warm and cold or fresh and salty, but copepods experience water as we would feel the texture of different fabrics. For them the contrast between the still, Zen-like vibe of the boundary layer stands out against the other parts of the sea like silk versus corduroy, allowing them to easily detect and cluster within a much narrower patch of ocean. So in addition to increasing the shelf life of the female's footprint, these sections of sea do what all good singles bars do: they concentrate the horny hordes. The more copepods that arrive within the boundary layer, the thicker the crowd, and the more likely each male is to find a trail and begin his sprinting pursuit.

But future generations of copepods may have trouble finding these trusted singles bars. As climate change progresses, ocean surface temperatures warm. On one hand, warmer water on top could strengthen the boundary layers, but it also may shift where those layers occur, how much oxygen exists (warmer water holds less oxygen), and the availability of food within each zone. On the other hand, warmer oceans fuel fiercer storms, which churn the surface and can disturb or completely erase the previously reliable layered landmarks of the open blue.

Even when the male finds the trail, it's not all smooth sailing. In some species, an elaborate copepod courtship dance ensues, with the couple swirling around to size each other up. Are they the right species? Is she a virgin? Studies show male copepods can detect and preferentially pursue unmated females, likely by sensing some kind of chemical cue. If the male decides she's the one, he lunges and hangs on tight.

The next ordeal in the quest for sex often involves a vigorous "rejection dance" by the pounced-upon female. Whether the extreme flips and violent shaking are a way for females to discourage fertilization when it is not needed or a way to test the male's mettle is not known.

The final act, the driving force behind all that seeking, happens when the male transfers and then cements good and tight a small packet of sperm, called a spermatophore, into one of two genital pores in the tail segment of the female. He does this by using his fifth pair of legs.

In an arctic species of copepod, males are not perfectly symmetrical; they have a preferred dominant leg for their sexy business, and the other leg (as well as their antennule) is used simply to hold on to the female. So it is that lefty males plug up the left pore and righty males go for the right one, leaving most females mated only on one side. Of course, having two genital openings means any female could mate again if she chooses in order to top off the other side—as long as the second lover used the opposite leg from her previous mate.

In contrast to the quiet waters of copepod singles bars, Nassau grouper go for an edgier scene. Solitary hunters, these grouper are the tigers of Caribbean coral reefs, growing to about three feet long and living for an average of sixteen years. Aggressive and highly territorial, they are not often seen together. Until the winter moons rise.

As the days shorten and temperatures drop, something awakens within these usually "homebody fish." For a few days each year, they go from hermit to hedonist: an uncontrollable itch drives them from their home reefs to travel up to a hundred miles or more to engage in a massive orgy. It's a remarkable journey cloaked in mystery. How do isolated individuals that barely stray more than the equivalent of a few blocks on a reef find a sex party that may occur all the way on the other side of an island—and find it *in time?* For Nassau grouper, the bacchanalia lasts for only two or three days each year.

Around Little Cayman Island, south of Cuba, researchers with the Grouper Moon Project have begun to unravel the mystery. Nassau grouper tagged with small acoustic transmitters can be heard by an array of underwater listening stations that encircle the island. The recorded "pings" of passing grouper paint a picture of how these lone fish seek each other out for their annual sex fest.

For many, it starts one or two days after the full moon, with a journey from their home turf to the outer boundary of the reef. There they hover, watching and waiting. And while they wait, some begin to slip into something a little more inviting.

Fish have a remarkable ability to change their coloration. Like birds, they often display bold, dramatic tones during mating season. In Nassau grouper, the switch is from a daily desert camouflage of mottled browns and beiges for more "black tie" attire: a sexy two-tone get-up that contrasts a bright white belly with a dark chocolate back. Others go for an all-brown look. Displaying these colors is one way they advertise their readiness to spawn. The darker and boldly contrasting shades seems to signal "friend with benefits" to other Nassau grouper—an important gesture by a normally highly territorial fish. Last thing anyone at an orgy wants is a swift fin-kick to the face for getting too friendly.

With their sexual intentions clearly on display, the fish watch for groups of other Nassau grouper, many similarly dressed, passing by along the border of the reef. They then swim out to join the party as it migrates toward the final destination. Some fish meander along the reef edge, swimming fair distances—even encircling the entire island—as they look for other fish and visit former breeding sites. Others simply swim out to the edge and wait for the caravan to come by them, slipping into the stream of fish when it does. The oldest and biggest fish tend to head for the site first, arriving earlier and staying longer than younger, smaller individuals.

What makes for the perfect love hotel for such traveling group spawners varies by species and region, but for most, a striking geologic formation often features in the mix: the edge of a steep drop-off or a large promontory that juts up or out into the open blue. These sites tend to have strong currents, which might help push newly fertilized eggs offshore and away from predators, or they may sweep the larvae in to safe habitat on the reef. There are several theories about what makes for a good spawning site, but it seems that location plus timing of the gathering combine to offer favorable conditions for newly conceived baby fish.

By two or three days after the full moon, the convoys of sexually aroused Nassau grouper swell in size as more and more hitchhiking fish join the group. By the fourth day, every single adult Nassau grouper from the waters surrounding Little Cayman can be found at the southwestern tip of the reef, an aggregation of four thousand fish that represents the entire breeding population of the island.

And that's where the problems lie. Migrating to the same spot year after year on a highly predictable schedule certainly helps fish find mates, but it also helps fishers find fish. Spawning aggregations offer an extremely lucrative fishing opportunity. There, in one small location, swim all the biggest fish from an otherwise scattered population. For fishers, it's like shooting fish in a barrel. On Little Cayman, when fishers first discovered a spawning aggregation in 2001, there were approximately seven thousand fish. Two years later, about three thousand Nassau grouper remained. A few fishers with simple hook-and-line gear removed more than half the total population.

Part of the reason why such enormous declines happened so fast has to do with the tenacity with which Nassau grouper seek out their sex parties. If Nassau grouper were more like people, then seeing a shrinking number of participants at the party might dampen the desire to join in a potential mass spawn of sperm and eggs (collectively called "gametes")—it takes a certain crowd for a place to feel inviting. As Dr. Brice Semmens, the lead researcher for the Grouper Moon Project, notes, "It's like showing up at a dive bar. There are a couple of snaggle-toothed individuals hanging around, but not the kind of folks you want to share your gametes with."

But Nassau grouper don't respond this way. Semmens and colleagues' research shows that as their numbers decline, Nassau grouper keep looking for fellow fish to follow; they keep showing up to the same spot on the reef, at the same time every year, and they stick around for even longer. The fewer fish in the group, the greater the amount of time the adults spend at the aggregation site.

Researchers still don't know why individuals hang around for extended periods as their populations shrink. Maybe it just takes longer for one or two of those particularly randy fish to show up and get the party started; or it could be that the cues to spawn depend on density, so the fish are waiting for their numbers to build. More data is needed to understand why they stick around and what spawning within these smaller aggregations looks like. What this means, however, is that remaining Nassau grouper will continue to seek out and form aggregations even as their numbers plummet, which provides fishers with ever-increasing and ample time to fish them out.

The good news is that the body of research on spawning aggregations is growing. With this knowledge in hand, governments can start making more informed decisions about the trade-offs between short-term income from catch and long-term benefits from healthy populations of fish.

SEARCH TACTIC NO. 2:
USE A LITTLE MAGNETIC ATTRACTION

Female stingrays have nowhere to hide. Try as they might to bury themselves in the sand, patrolling males on the hunt for love have a sixth sense for finding a mate: they can home in on her heartbeat.

The female's normal body activity, like all living things, emits signature electrical pulses. In round stingrays, the male's extremely sensitive electromagnetic sensory system is tuned to the precise frequency of the female's signal. Skimming the seafloor, the male senses her location, swings around into position above her, sweeps away the sand by flapping his wings, and, well, the rest is easy.

Sharks, as close cousins to rays, may similarly use electromagnetism to find mates over even greater distances. Take, for example, the hammerhead, which is known to gather in large schools around underwater mountains that rise up from the seafloor in the middle of the open ocean. Called seamounts, these features make attractive mating and feeding sites. The jagged slopes provide habitat and alter the surrounding water flow, creating upwelling and eddies that trap plankton and larvae—a banquet of food for anyone swimming by. These conditions allow seamounts to host a diversity of life far exceeding that of surrounding waters and make them great rest stops for far-ranging pelagic (open ocean) species to meet and mingle. Despite their dramatic relief, though, such seamounts are tiny within the giant bowl of an ocean basin. Hammerheads, like sea turtles and sea birds, likely find these hot spots by following an invisible map, courtesy of their ability to sense the magnetic fields of the earth.

Every year, sharks descend upon the El Bajo Espiritu Santo seamount in the Gulf of California. Forming a liquid wreath crowning the peak, their smoky silhouettes number in the dozens. These mostly female groups consist of larger, physically mature and fit females in the center, who push the smaller, younger females out to the perimeter. For a male hammerhead, such organization offers an easy way to find preferred mates: they head for the bull's-eye.

Dr. Peter Klimley, a shark expert who has tracked the movements of hammerheads for over two decades, has witnessed mature males dashing into the middle of the school to mate with the bigger females as they cruise along. The mating act rarely has been observed but when it happens, it's a slow-motion free fall. Midwater, the male wraps himself around the female and the two mate while sinking downward. The entwined pair fall head-first into the rugged reef, their signature hammer-shaped heads touching the mountaintop before the two sharks pull apart and swim off on their separate ways.

Despite years of study, researchers still don't know where the hammerheads come from or where they go when they leave the seamount for the season. But, through the use of more sophisticated tagging and tracking technologies, they are starting to understand the way sharks navigate around this particular spot in the sea.

Leaving the seamount at dusk for distant feeding grounds, the sharks travel with uncanny exactitude, nearly retracing the same route at dawn. They swim midwater: too deep for celestial cues, and too far above the bottom to see their way along. Geomagnetic fields surrounding the seamount, however, line up along the sharks' nightly migration corridors almost perfectly. Klimley thinks the magnetic fields serve as unseen highways, allowing the hammerheads to orient within their open blue world.

We lack such a fine-tuned internal compass, but thanks to the technology behind GPS and sonar, it's easy for people to find these seamounts too. They have become popular fishing sites for commercial fisheries over the last three to four decades. With so many species from far and wide aggregating at these sites, fishing seamounts is similar to fishing spawning aggregations: a few fishers can do a lot of damage, leaving the reproductive account of the seamount in the red.

SEARCH TACTIC NO. 3:
THERE'S NO PLACE LIKE HOME

Just imagine spending your entire adult life in the big city, meeting tons of attractive fellow singles day in and day out, year after year, but having to wait until just before you kick the bucket to have sex. And if that is not torture enough, when you finally get to do it, your only option is to go back to your hometown and lose your virginity with someone from your high school.

That's a salmon's sex life in a nutshell.

After spending years out at sea surrounded by thousands of potential mates, most males and females must return celibate to their home stream in order to reproduce—often the very stream where their own parents lost their virginity and where they too were born. They have to fight their way, defying current and, at times, gravity, to scale waterfalls and dams. They swim onward and upward to the cold shallows. There, in the clear waters, over a mosaic of pebbled riverbed, pairs of lovers—sometimes joined by eager passers-by—together let loose their gametes, a climactic ending to an epic journey. The adults literally spawn until they die.

For many species, such homecomings are the one and only chance males and females have to get busy. For salmon, the ritual is a once-in-a-lifetime event in a babbling brook; for northern elephant seals, it's an annual affair of raucous sex on the beach, likely the same beach where the cavorting adults themselves came into the world.

But showing up to the party doesn't guarantee sexual success. Male elephant seals must select and defend the "right" stretch of sand in order to get lucky . . . and that means finding a territory that passes muster with Nature's most scrutinizing customers: expectant moms.

Tethered to their terrestrial ancestry by newborn pups' need for dry land, female elephant seals must find safe spots to deliver their pups out of the water. After spending nearly eight months at sea, a female northern elephant seal begins her journey back toward the coast around New Year's. We don't know exactly how she navigates, but we do know her route is precise: one satellite-tagged female followed nearly identical migration paths in 2006 as she did in 1995. Although a few stragglers may change sites now and again, in general, once an elephant seal has picked a birthing beach—also known as a rookery—it stays there. And the vast majority pick the place they were born.

The males must journey even farther, traveling south from waters off Alaska to emerge from the cold, frothing surf in December. They have spent their time hunting dense aggregations of prey as deep as five thousand feet in orca- and great white–laden seas. The risk of attack is higher in these waters than in the offshore regions where the females feed, but there is far more prey available, something males desperately need. A single bull male consumes an estimated one to two hundred pounds of fish and squid *per day* in order to bulk up for the ensuing winter's battles on the beach. For the males, swimming a few thousand miles to the right stretch of coastline is but the first—and one of the easier—steps in the quest to find a mate.

Once the males reach the shore, these more than four-thousand-pound goliaths haul out onto the sand and enter into a modern-day Battle of the Titans. They use their bodies as battering rams to slam into one another in epic contests of strength and grit as they fight for rule over patches of precious sand.

Witnessing two bulls battle is akin to watching sumo wrestlers charge in the ring, only a seal weighs as much as a pickup truck and there are no rules of engagement. A mature male elephant seal's long fleshy nose, from which their common name derives, may reach an overhang of two feet and is the defining feature of a strapping male. They can inflate this proboscis, and often do, as they rear up on hind flippers to hurl the entire weight of their upper bodies into the head and neck of an opponent. Teeth bared, the two will slash at each other's eyes and throats.

Most combatants retreat to the edge of the beach, bruised and beaten. There they spend a few lonely months before heading back out to sea. A handful of champions, however, join an elite order: they are the beach masters, a rank that confers the opportunity to mate not with one female, but with a harem of up to one hundred of them. That sums up the sex life of a male elephant seal: all must battle for sex, most will die virgins.

For females, the story is a little different. They give birth within days of arriving on shore in early January, nurse for four weeks, and then, just around Valentine's Day, they become sexually available once again for all of two to three days. It's a precious but narrow window of opportunity for the males—the only chance they have to sow their seeds before the females head back out to sea. Having chosen the beach, the females clump up around one of the few large bulls that dominate the sands. Their quest to find a mate ends there. After that, it is merely a matter of surviving the act of intercourse—more on that later—and today, a changing environment.

The tie to land makes the elephant seals' reproductive strategy vulnerable to climate change, which drives fiercer storms, rising tides, and increased erosion of coastline. Coastal development and competition for space with another species that also covets prime seaside real estate—us—will make it difficult for elephant seal females to find alternative territories should their home beaches become unsuitable for breeding. At the same time, efforts to protect these once-endangered species are paying off: populations are growing along the entire coastline, which affords us the opportunity to rethink how we might share the coast with big wildlife.

Sea turtles are another species well known for homecomings full of reproductive flair. In most species, males and females live the majority of the time on distant feeding grounds, returning to favored breeding spots several weeks each year. The journey can be impressive, with both males and females crossing ocean basins to get back home again. As with hammerheads, it's likely that long-distance navigation relies in part on the subtle magnetic map of the earth and, at closer range, the smell of the beach. The olfactory, or smelling, sense in baby sea turtles is well developed, allowing them to be imprinted with the specific scent of the sand from the nest where they are born. Like a person who grows up next door to a chocolate factory, they can smell their way back home from miles away.

For male green sea turtles off Mexico's Pacific coast, the quest to find a mate simply requires returning home and hanging around the shallows just off the beach. Once in the general vicinity, the males don't have to do that much more searching: the females all come to them.

Dr. Peter Dutton, an expert in sea turtle reproduction, explains that for females, the journey requires a bit more planning. It may take months for a female to travel from feeding grounds back to the beach where she was born. Making such a crossing, and arriving with enough energy reserves to then be able to pump out hundreds of eggs, requires females to begin preparing for reproduction months, if not years, in advance. Most sea turtle mamas only breed once every few years for this very reason. When they have had enough time to rest and recoup, though, the female hormone cycle kicks into gear and begins the process of ovulation. It's during this time that the wanderlust arises, urging the female to venture forth and return once again to her place of birth.

After her long and carefully timed journey, she is met by a flotilla of lustful males blocking the way to shore. What then ensues is the sea turtle version of king of the mountain, as males clamor to scramble atop a female's shell and hold their ground against other competitors.

In populations such as the Pacific greens off Mexico, Dutton has seen the competition really heat up, as males bite, shove, and attempt to knock one another off as the female swims along. Each male attempts to be the one and only sperm donor for a female by blocking access to her until she heads to shore to lay her eggs. One way to do this is by digging his claws into her shoulders to secure a firm grasp. Such determination to monopolize a female often results in one male riding one female piggyback style for days, and in a few cases up to *several weeks*.

These impressive holds on the female may be one of the reasons why sea turtle eggs have a reputation for virility: the observation that the males can hang on for a long time has permeated the culture. Ironically, this coveted sexual stamina has led to large-scale declines in many sea turtle populations around the globe as nests are dug up and eggs sold as a form of natural Viagra. (Who knew a reputation for keeping it up for hours could be so detrimental to survival?) Just to be clear, there is no evidence whatsoever that sea turtle eggs convey any such potency. But in dark corners of seedy beach bars, men may still be found slurping down the slick white golf-ball-shaped eggs. In recent decades, bans on harvesting sea turtle eggs in many countries as well as beach patrols by volunteers has helped reduce the raids on nesting beaches. In one of the more creative initiatives, a Wildcoast ad campaign featuring supermodel and Playboy cover girl Dorismar put a spin on the aphrodisiac angle by creating billboards with an image of Dorismar lounging seductively below the phrase: *Mi hombre no necesita huevos de Tortuga*. Translation: "My man doesn't need turtle eggs." These efforts have helped reduce demand, but a market—legal and illegal—still exists in many countries.

With a long time until they reach maturity (over a decade in some species) and relatively low reproductive rates (in part because females do not breed every year), sea turtles can't withstand intensive harvesting pressure on adults or their eggs. But harvesting is only one of the challenges faced by sea turtles today. Another, more subtle threat also looms—that of warming temperatures. Sea turtles plan their reproductive cycle to coincide with just the right temperature window within which to lay their eggs. This is because the temperature of the sand determines the sex of the hatchling. It's as if instead of X and Y chromosomes, our sex was dictated by the weather. Babies born in summer would be girls, those in winter, boys.

As climate change continues to warm the land, the sex ratio of sea turtles is at risk, with the potential for fewer males to be produced. So far, this change has not yet been detected, but researchers are on the lookout. Meanwhile, physical barriers such as seawalls and jetties, meant to protect homes and preserve sand for tourists, can often destroy the habitats female turtles rely on for nesting. None of these challenges are insurmountable, however. Clever campaigns like Wildcoast's and other innovations are proving that sea turtle conservationists are as tenacious as male sea turtles in their refusal to let go of the cause. The results in some cases have been promising and provide hope for the future of sea turtles, despite the threats.

The urge to return home in order to reproduce also propels female sharks to journey back into their shallow natal lagoons to give birth to their own live young. This extraordinary ability to return to the exact location of their own birth was only recently discovered, thanks to a genetic technique that first revolutionized crime scenes and daily talk shows before being put to use for understanding shark biology.

Dr. Kevin Feldheim at the Field Museum of Chicago pioneered in sharks an approach that uses short repetitive patterns of DNA, known as microsatellites, to identify individual animals. These markers are distinct in each shark, just as they are among people. That's right. Feldheim's tests use the same technique that forensic scientists use to find suspects—the same science behind the paternity tests made famous by the Maury Povich show to expose the real baby daddy. Feldheim just uses the technique to look at baby mamas too.

With colleagues at the Bimini Biological Field Station in the Bahamas, Feldheim took DNA samples from pregnant females and newborn lemon sharks over a twenty-year period, starting in 1993. From 1996 to 1998, I volunteered to spend sleepless nights in the mosquito-infested lagoon catching newborn pups and taking tiny fin samples for the DNA work. That's where I first observed that some sharks are born with belly buttons, and that the little males show off their "maleness" from day one. But I digress . . .

The study shows that sharks have homing skills that rival salmon: six sharks tagged as babies in the lagoon in the 1990s returned as birthing mothers between 2008 and 2012. For most species of sharks and rays, we don't know how males and females find one another, how often they reproduce, and or how many pups they may have at a time. For the vast majority of the over four hundred species, we barely know the basics of their biology. Researchers still don't know where lemon sharks go to get it on, or how the males find the females. Unlike sea turtles, the females returning to the lagoon are already pregnant upon their approach, and, thus far, no blockades of lustful male lemon sharks have been caught near shore during the females' arrival or departure. Like so much about sharks, many of these basic sexual behaviors and processes remain a complete mystery.

However, knowing that at least some shark species may be using very specific pupping grounds provides important insight for shark management: these females are picky; any old habitat won't do. Though lemon sharks have access to similar lagoons in dozens of locations around the Bahamas, at least some females appear to make use of only one in their lifetime. Lose that one spot—say, to coastal development—and we risk losing these breeders from the population. These studies also show the power of genetics to elucidate behaviors in long-lived, difficult-to-study species—a power that's helping reveal far more than remarkable homing behaviors—stay tuned.

SEARCH TACTIC NO. 4:
GO FOR THE SMOOTH BARITONE

Male blue whales are getting their Barry White on. And if female blue whales are anything like most other mammals—from koalas to humans—going for a deeper baritone could prove a successful tactic for males looking to find and win over some mates.

The attraction to a deep male voice has its roots in the fact that females can judge something about a male's fitness—or at least his size—by the sound of his voice: bigger males can make lower-pitched sounds than smaller males. This instinctive attraction to a burly bass is so strong that women generally still prefer guys with a deep voice in our own species, despite the fact that modern culture has pretty much erased the link between bigger and better. In blue whales, the allure may be similar, and it could explain why male blue whales have been dropping to a lower key over the past few decades.

The theory is a work in progress, but it stems from intriguing data collected since the 1960s by naval submarine listening arrays, research cruises, and even seafloor seismometers. All show blue whale songs have dropped in pitch consistently over time, by a total of approximately 30 percent since the first recordings. The strangest part of the pattern is that the same adjustment happens even though blue whales from different regions sing different songs.

Like Professor Higgins in *My Fair Lady,* researchers can pinpoint where a whale is from based on the song he sings (as far as we know, only the males sing). Despite these independent populations all singing different songs, the same drop in pitch has occurred across nearly every population. Imagine if, around the world, carolers at the holidays all decided to sing two keys lower on the register each year, no matter the carols being sung and without the aid of a conductor to coordinate the drop. That's what blue whales are doing on a global scale. And researchers think it might have to do with how the population is recovering from the impact of whaling.

Blue whales escaped early whaling efforts of the eighteenth and nineteenth centuries mostly because they sink when killed. But in the late nineteenth century, the use of explosive harpoons, steam—then diesel—engines, and air compressors meant whalers could now keep whales afloat long enough to bring them on board for processing. During the twentieth century, an estimated 380,000 blue whales were killed, with estimates of more than 95 percent of some populations (such as in Antarctica) wiped out by the 1960s. When the international moratorium on commercial whaling went into effect in 1986, a blue whale's options for finding a mate were few and far between.

Unlike humpbacks, gray whales, and right whales that congregate on shallow breeding grounds, blue whales are loners, wandering the sea looking for high concentrations of food and potential mates. So even before whaling drove down the population, they needed a reliable form of long-distance communication to stay connected. Sound is the perfect solution. It travels extremely well in water, even better than it does through air. With sound, whales can paint an acoustical picture that potentially allows them to hunt, navigate, and find a mate across thousands of miles.

We achieve this kind of connection via online personal ads or chat rooms; blue whales do it with song. Singing at very low frequencies—a few tens of hertz—baleen whales, including blue whales, are the subwoofers of the sea, producing sounds that may reach clear across ocean basins. It remains difficult for researchers to know the distances over which whales actually communicate; what we do know is that a baleen whale singing off New England could potentially "post" to another whale swimming off Bermuda.

And here's where the low-pitched singing comes into play. Given a fixed volume of air in the lungs of a whale, different frequencies—the equivalent of pitch in the music world—can be produced with different amounts of energy. Higher frequencies have more energy than lower-frequency sounds. So even though low-frequency sounds can travel great distances, researchers speculate that there could be a trade-off between these low-intensity (quieter) sounds and higher-pitched songs that can carry more energy— think louder. Relatively small shifts upward in pitch may allow for a whale to sing louder songs that would carry farther across the ocean.

Back in the 1960s, the few lone survivors of the whaling era would have stood the greatest chance of locating fellow mates by calling as loudly as possible. So blue whales postwhaling may have had to increase their song pitch in order to turn up the volume a bit. These would be the songs first captured by the listening arrays—the songs of blue whales spaced far and wide.

As populations recovered over the next several decades, the dynamics of whale life changed. With more whales in the water, the distances between whales on average decreased, so blue whales may not have to reach as far to find other whales. Instead, if a male is trying to show off his size and status, it behooves him to drop to a lower bass to let other males and females hear how big he is.

Figuring out exactly what is going on in the lives of any cetacean—the whales, dolphins, and porpoises of the world—remains a huge challenge, though. As Dr. Phillip Clapham, lead of NOAA's National Marine Mammal Lab's Cetacean Assessment and Ecology Program, puts it: studying cetaceans is like studying lions through a perpetual fog, with glimpses of the tips of their heads or tails every twenty minutes or so. It may take years for scientists to accumulate population size and sex ratio information that a deer or squirrel biologist could get in a single season—or even sometimes in an afternoon.

As Clapham sums it up: "Anyone with any sense whatsoever doesn't study whales." This sentiment rings true especially when you consider the scales involved. True world travelers, many of the largest whales lap ocean basins as if swimming in a pool. In one instance, Clapham recalls the track of an endangered female gray whale that had been tagged by colleagues in its summer home near Russia's Sakhalin Island north of Japan. This whale swam clear across the Pacific to the breeding grounds of eastern Pacific gray whales: the waters and lagoons of coastal Mexico. She stayed for *a few hours* and then swam all the way up through the Bering Sea, back to Russia. It's like someone in LA flying to Paris for dinner, and then heading back—only the whales are using their own limbs for locomotion!

In every aspect, whales live life on a grand scale that is difficult for us to imagine. We've known for a long time that they cover great distances, but now it appears they may converse over those same expanses. This would mean they've been talking across the seas long before telephone or telegraph wires were ever invented, and certainly before online dating went international.

Blue whales aren't the only ones using big booming basses to advertise their whereabouts either. Sperm whales may also employ this tactic, though with a slightly different effect. Fierce hunters, sperm whales are capable of taking down colossal squid and fish in deep, dark depths. The males lead a solitary life, punctuated by fleeting moments of intimacy that hinge on finding a highly dispersed and regularly unavailable mate: a female sperm whale only reproduces about once every five years. Similar to blue whales, female sperm whales do not all congregate in one region to give birth. Instead, a male has to take his chances, cruising the open sea in the hope of stumbling upon small groups of females, one of which may be ready to breed—if he's lucky.

One way he can detect relatively nearby females is by listening. Sperm whales, both males and females, emit a kind of clicking sound to communicate and echolocate. It's a pretty constant chatter, especially when feeding. The male may home in on all that clicking and add his own special flourish to the mix. A large male sperm whale can do far more than just click. These big boys can BOOM.

Adult male sperm whales let loose the loudest biologically produced sound on Earth. What exactly the incredibly loud calls are for we don't know. Theories include attracting mates, repelling competitors, and even stunning prey. What we do know is that each "boom" says a lot about that male.

Dr. Ted Cranford, an expert in how marine mammals produce and hear sound, has shown that the loud booms of the sperm whale are produced by a pair of phonic lips located in their nose. To figure this out, Cranford had to create detailed models of a sperm whale head—not an easy logistical task, and one he achieved by commandeering CT scanners built for studying rockets. The elaborate maps Cranford produced reveal that a male sperm whale's head is basically a giant amphitheater. Sound is generated by the phonic lips located at the upper front of the head, near the almost right-angle curve of the sperm whale's iconic profile. This sound deflects to the back of the skull, where it bounces off at a slightly downward and forward angle, finally passing out through the flattened forehead into the sea. Just as with an echo, not all the sound is perfectly released—some of it bounces back and forth a few times within the skull. So the big boom, if listened to closely, is more of a cascading series of "BOOM BOM Bom bom." The time it takes for the sound to travel from the front to the back of the skull, and then back out the forehead is directly related to the length of the nose. So the interval between the successive echoes provides a good estimation of the size of the sperm whale's nose, and by association, the sperm whale. Even more than a low bass hints at the size of a blue whale, the reverb of a sperm whale's call says it all.

The sperm whale's nose is the largest on the planet, and it is packed full of fats and oils, which help to focus the sound. But that material is energetically expensive for the body to produce. Even more important, these resources are locked up—if a whale begins to starve, it can't metabolize this huge amount of energy. Building a nose that big with materials that can't be repurposed for anything else is a big investment and likely reflects the nose's importance for survival—as well as sexual success. The nose continues to grow throughout a male's lifetime and is disproportionately larger than in females, indicating that some sexual selection is likely at play. Whether a bigger nose wins over more females or wards off competitors by advertising might, big males with even bigger noses likely secure more mates and over the long haul produce more offspring. Which means sperm whales win the title for having the largest organ on the planet under sexual selection . . . by a nose.

Despite satellite tags, sonar, and submersibles, we still have no idea where the biggest animals on Earth go to have sex. There is something romantic in that mystery. But it makes it hard to help them out after we've knocked them back. Especially since total size of a population may not be all that matters for their recovery.

We know the payoffs for a large nose must be substantial, otherwise that huge investment wouldn't be worth it. Part of the reward likely comes in increased reproductive success: a bigger nose wins more babes and the opportunity to father more babies. It is not hard to imagine that females, evolved to prefer these bigger, dominant males, might not respond as readily to smaller, less-developed suitors. If this is the case, then mating success might drop even more than we anticipate based upon just looking at declines in population numbers alone. From fish to cetaceans, researchers are trying to understand how removal of the most attractive members of a species may impact reproduction in these more subtle ways.

The long-distance calls of the great whales of the world create an eerie, almost ethereal symphony of sound—some of it audible to us, much of it not. But recent advancements in technology, such as buoys equipped with acoustical listening stations, are helping researchers better comprehend the basin-scale conversations that are occurring. When a whale off Newfoundland may be communicating with a whale off Bermuda, it forces us to rethink the dynamics of whale dating games—and how human activities may be drowning out the conversation.

The oceans are far noisier today than they were in the past. Greater ship traffic, more oil and gas drilling offshore, increased naval activity that relies on sonar—all of these have contributed to a rising din throughout the global seas. And many of these machine-generated sounds occur right within the same frequency bands marine mammals tend to use for their own communication. Like static on the line, our droning ship propellers, air gun blasts, and sonar are creating an "acoustical smog" that has reduced the range over which many marine animals can find mates (as well as navigate, hunt, and socialize) by over 50 percent in some regions. Imagine if that happened to the sense we depend upon most. If our ability to see dropped by half, how would that change the way we moved or courted our crush? And the problem affects more than just whales. As we shall see, many male fish rely on sound to attract mates to their nests.

The battle over what constitutes reasonable use of sound, and its effects on marine life, is one that pits the forces of the military, fossil fuel giants, and shipping conglomerates against marine scientists who are just beginning to understand the direct and indirect effects of this modern phenomena of "sound pollution" on animals—including the success of their sex lives.

It doesn't always work this way. Sometimes we're not the ones making the racket.

When the sound got so loud it woke up half the neighborhood of West Seattle, the city knew it had to put a stop to the noise. But who was behind this acoustic devilry? Was it a rogue band of Buddhist monks chanting from the hilltops? A gang of teens wielding didgeridoos?

Turns out the intermittent yet incessant deep buzz that rocked the seaside shores came not from land but from sea. Specifically, it came from small, sexually aroused male plainfin midshipman fish. The males lure in the ladies from deeper waters by emitting a constant hum that the females can follow from the black waters to the light of the shallow rocky shore. With the development of Seattle's coastline, hulls of dockside boats amplify the soliciting males' droning call, creating a disturbance of the peace.

Small, elongate fish about the length of your forearm, with a broad head and froggish face, these normally offshore residents find their way to the rocky intertidal to breed in the late spring through summer. The males arrive first, staking out territory and building a nest in the shallows. Once settled into their nests, it's game on as they compete to see who can produce the sexiest signal: a low, deep vibration that sounds like a cross between a foghorn and a remote-controlled airplane.

During mating season, as her body prepares to lay eggs, hormones enhance the female fish's ability to hear the specific frequency of the male's hum. From the depths, she can search out a mate from miles away. Similar to how hormonal surges allow a new mother to cue into the faintest peep from her babe, the female plainfin midshipman fish's hearing changes, allowing her to hear the male's vibrato loud and clear. Conveniently, after she lays her eggs, the hormones shift back. She can now (blissfully?) tune out all the incessant buzzing of nearby courting males. Like a woman who eventually stops noticing the catcalls on her walk to work each day, the female plainfin midshipman swims in peace back to the deep. The residents of West Seattle aren't so fortunate. They have to tolerate the singing males, in full swing, through the end of summer.

SEARCH TACTIC NO. 5: MAKE A SPLASH

If you want to grow your orgy numbers, nothing beats a full-blown belly flop to attract the neighbors. That may be the strategy behind the extraordinary flights of mobula rays, stealth bomber–looking cousins of manta rays and sharks that momentarily defy gravity and their watery roots to burst forth from the sea in spectacular displays. Nearly all of the multiple species of mobulas seem to have this penchant for aerial acrobatics, leaping up to six feet out of the water before splashing down with a crack at the surface, but none do it in such extraordinary a fashion as the mobulas of the Gulf of California.

Normally solitary swimmers, these rays gather every year in seemingly endless numbers, their schools reaching up to a mile across. Looking down through the clear waters at a seasonal aggregation of hundreds of thousands of these rays is like watching a moving M. C. Escher painting: the diamond shapes of the mobulas create a deformed checkerboard that appears at least a dozen layers deep. With each layer moving at a slightly different speed, it is hard to know where to focus. Until they leap. Then it's all eyes on the flyer. Josh Stewart, a researcher with the Gulf of California program at Scripps Institution of Oceanography, likens the experience of sitting in a small boat floating atop a school of these rays to being inside a giant popcorn machine: the rays continuously "pop" out of the sea in all directions. No matter where you look, there are rays slicing upward, their winglike pectoral fins flapping through the air before flattening out to present a gleaming white belly that they pivot into a perfect parallel with the surface just prior to impact. When they hit broadside like this, the water splashes with a sharp smack, creating a loud sound that is likely the reason for these magnificent flights.

Why the mobulas jump is still up for debate, but it's a mystery Stewart and his team are beginning to unravel. Two of the original theories, that the jumps dislodge pesky parasites or are ways for males to show off to females, haven't landed well. As Stewart notes, rays are known for visits to "cleaning stations" where smaller fish eagerly pick off any parasites. And, he adds, "a parasite knocked off its host by a little jumping is a pretty shitty parasite." As for the idea that the leaps are flashy moves by males, well, the females leap just as often. Stewart's current theory is that the airborne escapades are a way of drawing a crowd. In other manta and mobula populations, this kind of leaping tends to happen when smaller groups form to group feed, attracting a few dozen other rays at a time. (It's unlikely the schools in the Gulf of California are feeding aggregations, though, as they are just too big. With several hundred thousand mouths in front, the rays in back wouldn't be left with much to eat.)

So the idea that slaps are a ray's way of communicating (they do not appear to make any other noises) sounds reasonable. The Gulf of California is a big place, and although seasonal cues may hint that the time is nigh for some group sex, individual mobulas need a way to find the school as it travels through the sea. Like a New Orleans parade that picks up a crowd as it weaves its way through the streets, the slap-downs of flying rays can be heard a long way off and may help mobulas that have come into the region find their fellow revelers. Once gathered together, it's far easier to find mates and even select the best of the bunch—which *could* be something the lively leaps help both males and females with too. Hearing the distinct crack, watching the exuberance with which the jumps unfold, it's hard not to think these rays are showing off, at least just a little.

THE QUEST TO FIND a mate presents a universal challenge to nearly every species on Earth, ours included. Our own quest may be as simple as buying a drink for an attractive guy or gal at the other end of the bar or as complex as a journey halfway around the world to find a spouse; it may look like a fancy debutante ball or perhaps a singles-only cruise. Under the sea, species deploy similarly diverse tactics to succeed in their own mate-finding missions.

Yet all of these work only to *identify* a potential mate, to close the gap that may exist between male and female. These approaches do little to ensure that upon contact, the sparks will fly. Ultimately, the seeking and the finding are but step one of a much longer process. The next chapter explores how marine life moves from this initial contact phase through to courtship . . . all while keeping an eye on the final goal of consummation.

2

LURING A LOVER

The Art of Salty Seduction

SEX-SEA TRIVIA

- *Pee can be a powerful love potion in the sea.*
- *Some male fish fake fatherhood to score more fertilizations.*
- *Cuttlefish are convincing cross-dressers.*
- *In some fish, the smaller the male, the bigger the cojones.*

SEX-SEA SOUNDTRACK

1. "Let's Get It On"—Marvin Gaye
2. "Love Potion #9"—The Clovers
(written by Jerry Leiber and Mike Stoller)
3. "It's Business Time"—Flight of the Conchords
4. "Lola"—The Kinks

She had been eyeing him for days now. Walking by his swanky house on the way to and from work. He was big, strong, and known for throwing his weight around the neighborhood. She liked his tough-guy nature. And yesterday she really fell hard.

It was the smell of him that did it. A light breeze was blowing, and as she passed by, his manly waft sent her head spinning. He was standing there on the porch, working out, sweat glistening off his flexing muscles. His scent was intoxicating. That's when she decided he was the one.

Her mind made up, she set off in the morning and sidled up his walkway and rang the bell. Showing up unannounced was risky. This guy was known for being cantankerous at best, violent at worst. But she was ready. As soon as the door opened, she let loose a stream of urine right on his doorstep . . . and then ran like hell. A few days more of that, she knew, and he'd be all hers.

WHILE A GOLDEN SHOWER is an unusual fetish for our species, the tactic proves a popular love potion in the animal kingdom, including under the sea. Such seduction-by-exotic-scent is only one of many tactics employed by marine life in their attempt to woo a mate. Just as we have our own diverse flirtation tool kit—the batting of long lashes, sultry salsa

moves on the dance floor—marine life also mixes it up when it comes
to seduction strategies, which run the gamut from sweet to scandalously
sneaky. Some species choose the burlesque route, shimmying and showing off a little skin—or scale—to gain the attention of a potential lover.
Others flash their status by building a big swanky nest or performing a
highly choreographed song and dance. Still others simply swap sex halfway
through their life cycle and take over the harem they once helped populate.
Different strokes work for different folks when setting the mood in the sea.
However, unlike our species, in which different individuals find different
forms of seduction alluring, under the sea, each species tends to stick to
one kind of approach or another.

This means that once we know how a few individuals court, we can be
fairly certain the same rules apply to the rest of that species. In other words,
male cod don't have to figure out if the bouquet of flowers or the pair of
handcuffs will work better to entice a mate—all female cod are likely to
find the same kind of thing arousing. That's good news for suitors, who
can stick to perfecting one mating dance, or building one kind of nest, or
concocting one perfect blend of pee. But this reliance on one technique can
also make a species vulnerable. Should human activities change the way
flirtation and courtship can happen, there are no alternative mating rituals
to fall back on. If the best performers disappear (say, plucked as preferred
specimens in the aquarium trade), there's no option for a male who is a
lousy dancer to attempt a romantic candlelit dinner for two instead. Having accomplished the quest to find a mate, individuals looking to seal the
deal must now successfully seduce—an art form that can enhance both
sexual success as well as vulnerability beneath the sea.

THE SCENT OF SEDUCTION

In the world of lobster sex, nothing says "let's get it on" like peeing in your
lover's face. Males and females rely on the sultry scent of their urine to set
the mood and, once engaged, keep potential rivals at bay until a couple
has completed their crustacean consummation. For anyone who has ever

tumbled through the surf and gotten a gallon of saltwater up their nose, the idea of smelling in the sea may sound a bit uncomfortable. But lobsters don't smell through a nose; they use their antennules, the smaller of a lobster's two sets of antennae.

When a lobster flicks these small rods through the water, it's sniffing around its environment, picking up subtle gradients in chemicals that hint at food, predators, or other lobsters. When a female Maine lobster, also known as an American lobster, goes to court a mate—the males lure the females in this species—she uses her antennules to smell her way to the male and then, in a twist, seduces him with her own intoxicating scent.

Maine lobsters—females and males—can say a lot with their pee. Males enter into veritable pissing contests as they do battle for the most coveted shelters—the ones just big enough for two lobsters to mate in style. The skirmishes start with each male shooting streams of urine at the other and progress to an underwater version of heavyweight boxing. Able to grow to more than forty-four pounds (twenty kilograms), Maine lobsters are one of the largest crustaceans in the world, and their fights can be intense. They bash, lock and crush claws, snap off legs, and snip off bits of antenna. When Mike Tyson bit off a chunk of Evander Holyfield's ear, he was boxing lobster-style.

But in the lobster world, there's rarely a rematch. Not right away at least. Boston University professor Dr. Jelle Atema has shown in experiments that the losing lobster can remember the smell of his vanquisher's pee and will refuse to engage in another brawl for at least a week. A winning lobster, on the other hand, seems to gain a bit of confidence with each conquest, an attitude that may be reflected in the smell of his urine.

In addition to increasing the volume of their pee, winning lobsters likely add a kind of signature to their scent, something that broadcasts their credentials and serves to intimidate rival males. Though he has yet to identify the specific compound, Atema notes that it's likely lobster urine goes beyond just identifying individuals; for dominant males, it may also say "I'm a badass, and I'm going to kick your shelled butt." Such an "eau de confidence" would be mighty attractive to a female lobster.

Drawn to aggressive males, a single female lobster faces a bit of a quandary: she needs to mate without being mauled. During mating season, big male lobsters are complete brutes. Typical behavior often involves nightly raids of neighboring lobster shelters, including those of females. Approaching another lobster's den, the big male will stop and fire a stream of urine in the front door. A few minutes later, the resident lobster—if it is wise—steps out of the den and makes way for the big guy to back his way in. The bully stays only for a few minutes, you know, just to show he can. He then moves on to repeat the harassment farther down the block.

It's the lobster version of chemical and psychological warfare. By continuously bullying his neighbors, a male never lets anyone forget who's in charge.

For a female, seducing a dominant male lobster is therefore a bit like trying to woo the Incredible Hulk when he's in a full rage. It demands a careful approach and a powerful love potion—strong enough to keep him entranced for up to *one week*.

Lobsters like their foreplay.

Lucky for the females, twenty-five million years have provided ample time to refine their skills as apothecaries. Arriving at the entrance of an aggressive male's shelter, all a female lobster needs to do is spritz him with some of her pee, a little each day over several days, and he will be putty in her claws.

The ability to shoot pee forward—something male and female lobsters both put to good use—requires some unique engineering (the mammalian penis is another example). For most animals, the outgoing waste ducts evolved to point *away* from the head, for obvious reasons. In lobsters, however, the bladder sits below the brain, with two reservoirs storing copious amounts of urine located just under their eyestalks. These connect to two nozzles through which both males and females can squirt pee. This powerful concoction then can be swept forward via the strong current created by the lobster's breathing. This technique allows their pee to shoot straight out in front of them . . . up to seven body lengths away. That's the equivalent of a teenager being able to pee from the back of a forty-foot school bus and hit the front windshield.

The female's strategy is to be coy and quick. She makes daily and brief visits to her heartthrob's den, flicking her antennules inside to smell him out and then spritzing him quickly in the face with her own pee (those front-facing nozzles come in handy) before getting the hell out of there. At first, each time the female returns, the male may lunge at her, even land a good swat or two. But eventually, her love potion begins to take hold.

The male starts to lets loose his own stream of urine, furiously fanning the fin-like appendages called swimmerettes located under a lobster's tail. This action draws her scent inward while flushing a mix of his and her urine out behind him. As most lobster shelters come with a back door for quick escapes, the mixed aroma of the his-and-her lobster pee wafts outward, broadcasting the lovers' intentions widely—lobsters may do it in dens, but they are far from discreet.

Eventually, the male calms down enough for the female to make her move and enter his den. She'll stay part time at first and only allow him to get to first or second base: there's lots of heavy petting with antennae and jointed limbs going on, but that's about it. Diane Cowan, senior scientist at The Lobster Conservancy and a former student of Atema's, explains that for females, this part-time moving-in together is a test. The female learns whether the male really controls the shelter or if another male can come kick him out. More than anything, a female lobster needs to know her partner can offer her total protection when it comes time to mate. For the male, it's a chance for him to learn if she is really ripe and ready—something he can likely smell *and* taste. As lobsters have the equivalent of taste buds on their legs, the constant touching between two courting lobsters is really more of a mutual tasting—they are licking each other with their feet. It's kinky stuff.

Like all arthropods, lobsters have their skeleton on the outside of their bodies. This means as they grow, they have to molt, shedding their old shell for newer, bigger ones to accommodate their progressively larger body. Although a female can mate between molts, mating just after molting is the preferred time. Here's why: when a female molts out of her too-tight bodice, she also sheds her personal sperm bank, a small receptacle located on the underside of her tail. This is where male lobsters deposit their sperm packs, allowing the female to draw down on the supply as her eggs develop and become ready for release. Any leftover sperm from past mating events are cast off along with the old pouch; a new, empty receptacle emerges with the new shell.

In other words, female lobsters can lose *and regain* their virginity.

For a male, mating with a freshly molted female offers the opportunity to fill that empty sperm pouch with stores of his—and only his—sperm. For a female, it means having the chance to fill up her new tank right away, allowing her to fertilize and brood a full batch (or two) of eggs before the next molt cycle—without having to mate again. There is only one problem: a just-molted female lobster is at her most vulnerable.

As if tossing off a suit of armor in exchange for a shimmery silk robe, a female lobster emerges from a molt delicately attired—and unable to stand. Her soft new shell will take at least thirty minutes to harden enough for her legs to support her own weight; it will be several days before it serves as effective body armor again. To mate at this stage a female lobster is at the mercy of the male—a large, strong, extremely aggressive, giant-clawed individual.

That's where her love potion comes into such great effect. The scent of a pre-molt female is the ultimate aphrodisiac for a male. Once the male has melted into a more hospitable nature and the female determines he does indeed lord over this shelter, she'll move in full time.

For the next several days, the two will snuggle up in the same shelter, leaving to hunt and go about their lobster business, but returning to the same "home." Then the time comes for the final act, and it's now, more than ever, that the female must ensure the male stays rather subdued.

For nearly the entire time that she lives with the male, the female is at his side or behind him. But now, in these final moments, she circles around to face him, eyestalks to eyestalks. He spreads his claws wider and down, almost as if bowing before her.

And then, she knights him.

Standing before the male, the female solemnly lifts her claw and taps him on the shoulder, then repeats the movement on the opposite shoulder. It's a signal that seems to convey a message: Don't leave me now. Standing face to face, the two also partake in a massive, mutual golden shower. She then walks to the back of the shelter and strips.

It may take up to an hour or more for her to molt, but exactly thirty minutes after she kicks off the final bit of shell, it's time to get down to business. In lobsters, the actual act of copulation is a surprisingly romantic—albeit swift—affair. Under her spell, the former tyrant of a male is transformed into a gentle lover. Immediately after her molt he stands guard over her soft body, resting on closed claws and may even lightly stroke her with his antennae. At the appointed time, the male circles behind her, assuming a doggy-style mount. But then, in what may be the most tender act of lovemaking in the invertebrate kingdom, he lifts her gently off the seafloor and cradles her in his small walking legs.

He braces himself, with big front claws and tail pressing into the sand, and gently turns her onto her back, pulling her up toward him. She assists by stretching out her tail to lay as flat as possible. Belly to belly, they then fan their swimmerettes vigorously as he inserts the first pair of modified swimmerettes, called gonopods, into her sperm receptacle. Each gonopod is a half tube that he squeezes together to form a hollow rod through which the spermatophore is passed. She hangs there, in the hammock of his arms—er, legs—as he completes several thrusting motions. There is more mutual fanning and urinating, and then he gently rolls her back over and sets her down. Exhausted, she returns to the back of the shelter. A few days later she'll move out . . . and another female will move in.

Maine lobsters are serial monogamists.

The combined scent of a courting couple, beyond subduing the male, may also help control the mobs of females lining up outside the entrance of a dominant male's den. The male's urine on its own may call "Come hither," but mixed with a female's, it seems to say "currently occupied."

As has been shown in other animals—including people—the timing of a female's cycle can be affected by cues from other females or males in the area. Thus, we get college dorms where all the women menstruate at the same time each month and female lobsters who can stagger their molts so that each has an opportunity to mate with the dominant male at the most opportune time. The specific mechanism of how this works is still unknown, but chemical cues are likely at play.

In common slipper shell snails, females take the transformative power of pheromones even one step further. Found along the rocky shores of New England, slipper shells, also known as boat shells, are mollusks with an unusual habit of stacking one on top of another in tall vertical columns that resemble the Leaning Tower of Pisa. They have deeply domed backs that, when flipped over, resemble a fat canoe. This underside contains an inner shelf that extends about halfway across the opening, making it look like a slipper. After a few weeks floating through the open sea as a tiny veliger (baby snail), slipper shells head for the seafloor, and that's when strange things start to happen.

If a baby snail touches bottom and is far from any other slipper shells, it will rapidly mature as a male first, and then, very quickly, turn into a female. Slipper shells are sex changers, known in the science world as sequential hermaphrodites. The word *hermaphrodite* is a combination of the male god Hermes with the female goddess Aphrodite. The term most commonly refers to an individual having the genitalia of both sexes. Such *simultaneous hermaphrodites* are common in marine species as well as plants—for example, roses have male and female parts in a single flower. But slipper shells are a different kind of hermaphrodite—a sequential one—which means an individual develops first as one sex, then another. This sexual strategy is pretty widespread in the sea and something we will dive into in the next chapter in greater detail. For now, consider the lone slipper shell's ability to first grow a penis, then reabsorb it, sometimes within less than two months, as it moves from male into female existence. In mature female form, an isolated slipper shell then emits a strong pheromone to attract as many other snails to her as possible.

The trick though, is that she isn't attracting males to her. She is attracting juveniles—then coaxing them to persist as males for much longer than they would if on their own. Typically, juveniles beeline toward already-existing stacks, following the female's tantalizing chemical trail, and drop down to latch onto other snails. With the help of the female's strong scent, fresh-from-the-surface snails arrive to develop as males, each one unfurling a long, extendable penis that reaches down to fertilize the anchoring female. Building stacks up to half a dozen individuals in height, a slipper shell male is rather well endowed. All snails in a tower except the first female forgo their sex change, instead remaining male and fertilizing the enormous numbers of eggs the female produces. Once a stack has enough males pumping out enough sperm to fertilize the female's eggs, the male closest to the bottom female may then transition into another female. Rather than compete with other males to fertilize a limited set of eggs, this individual switches to female and starts producing eggs of its own to take advantage of the surplus of sperm. Of course, as a newly transitioned female, the second-floor snail will start emitting pheromones along with the first female, attracting more juveniles to build an ever-taller skyscraper of lovers. No wonder their scientific name is *Crepidula fornicata*.

This is the power of pee—or potentially any chemical signal—in the sea. After decades of study, Atema feels he has only scratched the surface. There is "an unlimited possibility" of subtle cues with which animals can communicate through chemical signaling. But the messaging only works if animals can detect the smell within their watery environment. And that depends on two things, both of which we may be disrupting.

First, chemical cues are carried by tiny molecules that interact with the seawater they drift through. Change the chemistry of the water, say, by slightly lowering the pH, and the structure of the signal molecules change. Like rearranging the letters on a notecard, the message no longer makes any sense; alternatively, a drop in pH could interfere with the receptor cells on an animal, like those covering lobster antennules, reducing their capacity to "read" the message. Right now, our continued consumption of fossil fuel emits enormous amounts of carbon dioxide into the air every day, much of which the oceans absorb. When carbon dioxide and seawater mix, a series of chemical reactions occur that make the seawater more acidic (a lower pH). Known as ocean acidification, this effect of climate change is the evil twin of global warming in the sea, and it could have major consequences for the way animals send and receive messages there.

Second, like the kid who stinks up the cafeteria by bringing a tuna fish sandwich to school, we are constantly adding strong smells to the sea. When Atema first began his work on lobsters, it was to discern the sublethal effects of oil spills on lobster health. One of the things he discovered is how well a foreign substance could mimic other pheromones, falsely attracting lobsters to a "food source" that contained no food. In this case, it was bricks soaked in kerosene. Lobsters would flock to and incessantly lick the bricks; a few days later, they fell ill, refusing to eat for up to a week at a time.

In other cases, pollutants may cover up naturally occurring odors, masking the signals animals rely on for communication. Together these changes risk disrupting the elaborate courtship and seduction of myriad species that depend upon a little aromatherapy to set the mood for sex. Fostering the right chemistry for sex in the sea requires addressing both local and global threats to the ocean's chemical balance—an enormous challenge but not one beyond reach.

THE SEX APPEAL OF A DUTIFUL DAD

We've seen it happen before. An attractive man, well dressed and polite, draws the attention of some single ladies; an attractive man, well dressed and polite and rocking a baby in his arms melts the heart of nearly every female in sight—single or otherwise. Some female fish feel the same way.

Especially seahorses, which are a group of sexual extremists like no other. First, they are relatively monogamous. These days, the power of paternity tests has shown that very few species are *really* monogamous. For a mating season or for life, promiscuity reigns. But a few species of seahorses appear impressively faithful to their partners, at least throughout a breeding season, even waiting for a partner to recover from injury or illness rather than pursue a new mate. The time frame of such commitment may last well over one hundred days—it's not a lifetime, but still, it's an anomaly for the animal kingdom, above and below the surface.

In addition to this unusual behavior, male seahorses take the dutiful dad thing to an extreme as well, going so far as to become pregnant. You read that correctly. In seahorses, the male, not the female, fertilizes the eggs and then nourishes the developing embryos inside a warm pouch that bathes the eggs in fluid and protects them until they hatch. The male then births live young that spill forth into the sea as perfect mini seahorses. Pregnant. Males.

But before *that* can happen, a male must convince a female to entrust him with her costly eggs. On the other hand, a female needs to ensure she can rely on the male to turn up when her eggs are ready. A female can only carry ripe eggs for about two or three days before she has to dump them—if the male is a no-show, that's an enormous waste of energy for the female.

In order to coordinate their love affair, males and females engage in a little foreplay each and every day over the five- to six-month breeding season. Cryptic as they are, our knowledge of the approximately forty species of seahorses in the world is very limited. But we've successfully spied on a few, and their morning courtship ritual looks a bit like a ballroom dance scene out of a Jane Austen novel.

A male might approach a potential mate with his head bowed down, ever so courteously, and flutter his fins rapidly. He also dilates the opening on his belly where the eggs are stored, inflating it at much as he can to show off his wares. He may also lighten the color of his stomach to accentuate the pouch. If the female likes what she sees, she'll reciprocate by similarly lightening in color and bowing her own head in return. The male and female, once met, then separate for the night, like good, chaste Victorian singles should.

For the next several months, at dawn each day, the female slowly swims over to the male's territory, where he patiently awaits her in their usual greeting spot. Upon her approach they both brighten in color, as if blushing. Then they each wrap their tails around a blade of sea grass and begin to circle their holdfast, the male swimming the larger outer sweep with the female on the inside. After a few moments, they release the stalk and line up side by side, then drift in parallel over to another blade of sea grass. As they do so, the male often wraps his tail around the female's, like two lovers holding hands on a morning stroll. This flirtatious dance repeats a few times, with the entire greeting lasting several minutes.

When a couple first gets together, it can take a few days for the female to finish developing her eggs and make the transfer to the male. After that, the brief, daily dance seems to be enough to help her gauge his stage of pregnancy and adjust the timing of her next batch of eggs accordingly. In seahorses, sex follows birth almost immediately.

Unlike most females, who require a little recuperation after giving birth to live young, male seahorses are ready to rock shortly after releasing the last of their babes into the wild. That's when the daily greeting turns into a full-blown seduction. The male usually gives birth at night, and the next morning, he indicates to the female that he's ready to move beyond dancing by displaying some energetic thrusting. He alternates between folding in half, contorting his tail up toward his head, and lengthening out. This forces water in and out of his belly pouch, helping it swell in size. Females find a big bulging abdomen rather hot, and after watching this display, she'll respond by pointing her snout upward. It's the equivalent of the subtle head nod toward the bedroom that says, hey, how about we go up *there?*

The courtship can last a long time—up to nine hours in one seahorse species. Then, after weeks of daily dancing and hours of foreplay, the pair swim upward in the water column and have sex for about five seconds. In this regard, female seahorses really are just like the majority of males in the animal kingdom. Where seahorses differ, however, has major implications for how their populations respond to fishing pressure—which is not well.

Dried seahorses are still sold as popular curios, ground up for traditional medicine, and consumed as an aphrodisiac—ironic, given that their greetings are so innocent and their sex bouts so fleeting. Additional harvests also occur for the live aquarium trade, despite the fact that seahorses are extremely difficult to keep alive due to their diet of live fish and a tendency to simply stress out. Over the past several decades, fishing to satisfy these markets has taken a toll on seahorse populations. The damage is more than may have been anticipated in part because of the way they reproduce.

Unlike fish that spawn directly into the water or a nest, seahorses are limited by how many eggs the male can carry at any one time. In general, this is from several dozen to perhaps a hundred. Over the course of a six-month breeding season, assuming they have sex every two weeks or so, seahorses pump out perhaps one or two thousand babies. That is tiny compared to the millions of eggs and sperm released by, say, cod. This is the first reason seahorse populations decline quickly when fished: they simply don't produce enough babies to replace the number harvested by fishers.

Second, while romantic, the strong allegiance to a mate in some species means that, should a male seahorse be scooped up, the female may continue to return to her mate's territory for days before looking for another male; even then, she likely won't be ready to mate right away, as her cycle is timed with her former mate's pregnancy. This means remaining females and males may miss out on breeding opportunities. The same thing happens if a female is taken, by the way, especially if the male is pregnant: he has to wait to give birth and then start the courtship anew, which may require several days for the female's egg cycle to get in synch.

Fewer babies and fewer mating opportunities add up to rapid declines under even low fishing pressure, something we couldn't have accounted for two or three decades ago. But we can now. As research provides more details of the courtship and reproduction of seahorses, we can set more realistic limits to harvest. In addition, the science on captive breeding of seahorses continues to improve every year, with major programs now successfully producing adorable, itty-bitty baby seahorses with which to supply the aquarium and medicinal trade demands and alleviate pressure on wild populations. These are steps in the right direction for ensuring the survival of the planet's only pregnant males.

Though none are as extreme as seahorses, there are many other fish species with male-dominated parental care. In some, males carry the fertilized eggs around on their bodies, or in their mouths. Many spend their energy building and guarding nests. For these species, having stay-at-home dads to protect and nurture their eggs frees up the females to go feed and store up on nutrients. Dr. Bob Warner of University of California, Santa Barbara explains that this is one of the benefits external fertilization bestows on females: when only one parent is needed to tend the eggs, the females can lay them and head out the door. The male, having to fertilize the eggs after they are laid, is left "holding the bag"—if he wants his off-spring to survive, he's got to stick around and protect the vulnerable eggs from predators. Guarding a nest also can help a male ensure that it is he, and not another male, who fertilizes the eggs the female lays in his nest. At least, that's the idea. It's not always such a successful strategy.

Consider the peacock wrasse, a roughly foot-long fish with beautiful turquoise and magenta accents that builds a nest of squishy seaweed. In this species, females don't seem impressed by male size, vigor of displays, or even how the nest is built. For them, the best dads—and thus the most attractive mates—are the ones that already have offspring, or at least off-spring-to-be. In other words, a male with eggs in his nest, like that single guy rocking the baby, is all she needs to see.

There are several reasons why such attraction makes sense from the female's side. First, a male guarding lots of eggs is unlikely to abandon the nest—he's got too much invested. Second, the presence of eggs shows that he's able to defend his nest—otherwise, all those eggs wouldn't still be there. Finally, and it's a little bleak but true—nest-guarding males are known for munching on their own eggs when hunger strikes. If a male does decide to snack on his own eggs, a female who mixes her eggs in with a bunch of others at least spreads out the cannibalism.

From the male's perspective, this whole "best dads are already dads" thing is a bit of a Catch-22. Each year or breeding cycle, every male starts with zero eggs. Imagine if single women all found divorcés with children from previous marriages to be way more attractive than men who hadn't yet sown their seeds. How would a single guy without kids ever manage to woo a woman to have kids with in the first place?

In several species, females give a guy a chance by placing only a few eggs in a male's nest at first, and then waiting to see how he fares. It's probationary parenthood, with the female checking back in a few hours or perhaps a day or two. If the male tends the nest well, he's rewarded with more eggs. But in the peacock wrasse, scientists still don't know how some males manage to convince females to lay those first few eggs. What we do know is that once a few males have succeeded, other males can benefit—at the successful dad's expense.

When it comes to proving fatherhood, some males just fake it.

Large "pirate" males in this species of wrasse regularly take over the nest of a successful—but smaller—neighbor. These bullies kick out the original dad and then dupe females into thinking they have invested the effort and built the nest themselves. (I suppose since females don't stick around long enough to see the eggs hatch, they have no idea the larvae look nothing like their father . . .) New females come and lay eggs that the con artist male then fertilizes. But the ruse doesn't end there.

Log enough hours watching wrasse reproduce, as Warner and his then–graduate student Dr. Eric van den Berghe did, and you will also see these pirate males take the deception a step further. Having spawned with lots of females in the usurped location, the male then ditches the nest. He can do this because out of the wings swoops in the original, dutiful dad to save the day—and the nest full of eggs. Since there is no way to tease apart which eggs belong to whom, the original dad has to care for the whole lot if he wants any of his own eggs to survive. The cheating male's deception buys him a nest, favor with the females, and a free babysitter for his eggs while he goes off to con more females using yet another male's nest. Sneaky wrassetard.

THE ALLURE OF AN ATTRACTIVE LOVE NEST

In many species a male can convince a female to lay her eggs in his territory, as opposed to the other guy's down the block, if he can construct a flashy nest. And no one builds as spectacular a nest as the male white-spotted pufferfish. (Bowerbird fans, I'm sure, will argue with me on this, but I hold my ground.)

Viewed from above, the nest of this puffer looks like a carefully constructed mandala, stretching up to six feet across. That in itself would be impressive, given the fish can fit in the palm of your hand. What's even more amazing is that the male sculpts these giant structures mostly by shaking his butt.

With a bright white underside and a golden back dotted with faint, cream-colored spots, this puffer was confirmed by scientists as a new species in 2014. Finding a new species in the ocean isn't all that surprising. Finding a new species that builds elaborate underwater structures that rival those of famous landscape architect Frederick Law Olmsted or the most radical design of modernist Antoni Gaudí, well, that's not so common.

Next time you visit the beach, imagine building a giant sand wheel with radiating spokes and troughs that stretch over one hundred feet across. Now imagine carving the whole thing by scooting around on your bottom. That's close to what these males must do to build their intricate love nests.

Swooping in low over the sand, the male vibrates his anal fin and wiggles his tail rapidly, driving his belly into the seafloor like a snowplow pushing through drifts. He alternates digging deep and gliding over areas, creating undulating valleys and ridges. After a week of solid work, he puts on the finishing touch: a decorative cornice, made of coral fragments and bits of shell, along the ridges. Then, he tidies up and moves to the central region, where he builds a mandala within the mandala out of the powdery, finest sand particles.

The male's elaborate nest rises slightly up from the monotonous sand plains like an ancient temple, visible from afar. The contrast of light and shadow likely catch a female's eye as she cruises the dimly lit sand plains. As she approaches, the male fans the central disc vigorously, shooting plumes of the white smoky sand into the air.

If intrigued, she enters the nest, and the male quickly circles her at a wide berth before darting toward and away from her several times. The male, having put out an enormous effort building this structure, can but hope that the sand—or whatever she's judging him on (we still don't really know)—suits her. If it does, she will move into the center. The male then makes a series of rapid dashes up and back along the various ridges of the nest line, as if to say, "Did you see this valley? How deep and wide? Or how about these peaks over here?" After a few such enthusiastic sprints, the male joins her in the center, often biting onto her cheek and pulling her alongside him as they spawn. Courtship over, eggs fertilized, the female takes off. The male remains behind, guarding the eggs in the center of the nest as he watches his fortress dissolve back into the sandy plains. When the baby puffers hatch a few days later, the male then abandons this nest to begin anew somewhere else.

Not all female fish go for such artistic types in looking for a decent guy to mate with. Many really just want a male that can defend the nest from egg-snatching predators and won't eat the tasty, fatty, bite-sized morsels himself. Bigger males with more fat reserves meet both of these concerns: they can better fight off predators and are less likely to ditch the nest to go forage or snack on the eggs themselves. For many species, seduction really comes down to size: bigger males generally convince more females to drop more eggs in their nests.

"Bigger" doesn't always mean total body size, though. In some species, females find a healthy set of anal glands a major turn-on. Anal glands may not sound all that sexy, but these sacs can carry potent antibiotics that help protect developing eggs from a whole host of infections. Males with larger glands can likely spread around more antibiotics. They lack the typical tackle of a male mammal, but that doesn't stop these male fish from showing off a manly pair to impress the ladies.

Bigger males also tend to build (or win) the best nest sites. What constitutes "best," however, depends on the species. Dr. Phil Hastings, professor of marine biology at Scripps Institution of Oceanography, explains that blennies are a case in point. These small, elongate fish dart and scoot around the bottom of the sea—be it rock, reef, or sandy floor—often poking their heads out of small crevices. With large eyes and a blunt head, they have a perpetual look of wonder—or constant alarm. Out of the nearly nine hundred species of blennies found around the world, nearly all of them lay eggs in some sort of nest that males then guard.

Across these hundreds of species, there is a wide variety of nest types, with males showing off all kinds of architectural prowess in order to woo a mate. In the redlip blenny, females prefer McMansions—large, roomy nests in which to lay their eggs. But big nests, like big houses, are costly to maintain. They may not carry a mortgage, but males pay the price of large nests in sweat equity. There's the constant work to remove pebbles that accumulate in the crevices, ongoing scraping to clean off the seaweed, and endless chasing of other fish from the territory. A smaller nest is far less of a headache, but it may not attract the females.

Male redlip blennies navigate this trade-off by assessing their sexual success throughout the season. He begins by tidying up one of the smaller nest sites within his territory. When a female arrives, she will size up a male as he returns the scrutiny. Males also prefer larger mates because bigger females can put more eggs in the nest—if he is going to sit there all day fanning eggs and chasing away predators, he might as well be doing it for the biggest pile of potential offspring the nest can hold. If they pass one another's muster, the female will then head into the shelter to inspect its dimensions. If she likes it, she'll lay her eggs and then depart. If she finds it too cramped, she'll leave, taking her eggs with her.

After one or two breeding sessions, a male redlip blenny takes stock of his breeding success: if the number of females laying in his nest seems low, he'll switch nests—nearly always upgrading to one with a more expansive interior—think vaulted ceilings and great flow. It's a level of self-evaluation surprising for a fish, but it helps the male maximize the trade-off between nest size and attraction. More often than not, the switch secures more mates.

But not all females desire big, airy abodes for their nest-laying; according to Hastings, some just want their nest to be *clean*.

Tube blennies go for the reclaimed "loft" apartments of the reef, living inside abandoned barnacle shells, called tests, or worm tubes. Although there is only so much one can do to refurbish an old building, Hastings's research has found females won't compromise when it comes to cleanliness. A female tube blenny simply won't drop her eggs in a dirty apartment. Courting males in these species seduce via sanitation.

This results in some interesting twists on the whole bigger-is-better preference. Already in short supply, big tubes tend to be older and more overgrown with algae and encrusting invertebrates. Sometimes smaller males, with cleaner tubes, wind up winning more eggs than bigger males, whose dirty, smelly bachelor pads don't pass the tube blenny equivalent of the white glove test.

This kind of picky taste may come with trade-offs, though. A choosy female might secure a better abode for hatching her babes, but it also means these blennies depend greatly on their tube-building buddies, the barnacles and worms. Changes to their populations can impact the blennies hard. The disappearance of Hastings's tube blennies from his original study site is an interesting case in point. The shallow reef that once was home to a city of tube blennies, all poking their heads out of their reclaimed tunnels, is now a bay thick with sediment. It's an all-too-familiar story: development along the coast led to increased runoff, which likely choked out many of the species that formerly lived there. Even though blennies might not have been directly affected, the increased sedimentation would have hit filter-feeding barnacles and worms fairly hard. If these populations died out, the tube blennies would then have lost their hideouts from predators as well as their supply of new apartments. Given the high standards of fastidious female tube blennies, the habitat may simply have run out of acceptable homes. It's impossible to know the exact sequence of events, but this kind of indirect effect is one worth considering—and it highlights two of the hardest aspects of managing marine resources: they sit downstream of all our waste, and the impacts may not always be direct, which can make them difficult to detect or anticipate.

FAN DANCE FLIRTATION

Give a six-year-old (or someone high on acid) a box of the brightest crayons imaginable and ask them to draw a fish, and you might get something that looks close to the splendid mandarinfish. Bright orange paisley swirls pasted against an electric-blue body also streak across the bright teal-colored face. This cacophony of color is all packed within a fish the size of a man's thumb. Wide, rounded pelvic fins hang down like two Ping-Pong paddles while equally broad pectoral fins flutter at the sides. As elaborately designed as these fins may be, the crowning jewel of the male is his sail-like dorsal fin.

A male can swiftly erect a stiff spine at the front of the dorsal fin in aggressive displays to ward off other males or in courtship to impress nearby females. As he does so, the spine pulls up the first section of the dorsal fin like hoisting a sail on a mast. Further back, he stiffens a broad second dorsal fin that when straightened nearly doubles the "height" of the fish. So as not to be unbalanced, the male can also extend the keel-like anal fin, which hangs like a rounded curtain from beneath his belly back toward the tail. For a finishing touch, he spreads his pectoral fins wide, as if flashing "jazz hands."

The full display expands the male's height profile significantly, making him look large and definitely in charge. Big males are quick to chase off any smaller males that invade their space. Should an equally large male approach, a dueling display may occur with each male rapidly flicking his dorsal fins, fluttering pelvic fins, and otherwise trying to out-compete the other. Watching two males circle one another is like witnessing an elaborate fan dance, as the competitors splay open and fold their psychedelic fins.

When it comes time to woo a female, males put on a similar show, adding a full-body vibration to the mix. They flutter their pectoral fins rapidly, creating a shimmering effect as they waggle their bodies side to side.

Females are choosy about whom they partner with, preferring the biggest males around; the largest females almost never spawn with males smaller than themselves unless they have no other choice. As bigger males have proportionately larger dorsal fin spines, it is likely that females are attracted to both the total size of the male and the length of his shaft. And for good reason—through the crafty use of their enlarged anal fin, bigger males can "satisfy" a large female's needs better than a smaller male can.

As a male and female mandarinfish get friendly, they line up alongside one another, matching up their vents (the opening on the underside of their body near the tail through which sperm or eggs escape). In a rather dexterous display, the male then curls his long anal fin up toward the female, creating a sort of funnel that keeps her eggs from dispersing too rapidly and gives him a greater chance of fertilizing the batch of a few hundred eggs before they drift off. Smaller males have smaller anal fins and likely cannot corral the eggs as effectively, which may lower fertilization rates.

Female mandarinfish aren't the only ones that prefer big flashy males. We like to ogle their Dr. Seuss-like colors too. In the aquarium trade, the biggest males, with the biggest dorsal fins, fetch the highest price in the market. Currently, all mandarinfish sold are collected from the wild, and nearly all fish in the marketplace are large males. This preferential selection is a double-whammy for a female: it reduces her chance of finding a decent mate, and if she does find one, the chances are sex will result in fewer offspring. Both of these consequences mean remaining mandarinfish will have a hard time producing enough new babies to replace the big males taken off the reef. One of the solutions to this problem would be to crack the code of how to breed these beautiful fish and provide a sustainable farmed source to supply the aquarium trade. Although hobbyists have managed to easily breed and raise them in captivity, so far preliminary efforts to produce the fish at a commercial scale have failed. Until alternative sources are available, improved management is needed to rein in the assault on the most attractive suitors in the population and rebalance sexual success in the population.

The fan-dance routine is common in many species of fish where males attempt to accentuate their body size by showing off some elaborate extensions. Tube blennies similarly use a fin flick here and a shimmy there to attract a female over to their abode. In contrast to multicolored mandarinfish, Hastings notes, blenny males may go for a monochrome look when seducing a mate. Deepening their coloration to almost black helps these males "pop" against the turquoise water and white sandy bottom. Once dark and sultry, they begin their dance, which looks a bit like a burlesque number crossed with some semaphore signaling.

In a typical sequence, the male pokes out of his tube and lifts his front-most dorsal fin spine. Often, the fully unfurled dorsal fin has a big conspicuous circle or spot on it, something that works to catch the female's eye. He'll also flare his gills, giving his head a much wider, rounder look. Then he goes straight back into his tube, as if he's playing a flirtatious game of peek-a-boo.

As these blennies live right next to one another, the competition to catch the eye of a female passing by can be pretty thick. There are lots of fights, something females likely use to judge newly arrived males in the neighborhood. According to Hastings, "It is as if the females are waiting to see if this new kid on the block can hold his own." Once he's scored some eggs, a male tends to simmer down a bit. You might think his aggressiveness is tempered by a new sense of responsibility (OK, instinct) to care for his developing offspring. Perhaps. But it could also be that such flirtation is dangerous, attracting attention from predators as much as females. There are trade-offs to being showy when you are but a small fish on a big reef.

SONG AND DANCE

Unlike tube blennies, haddock and cod never seem to tire of the opportunity to seduce a female, which they do by adding a little song to their dance. This is the foreplay behind your fish and chips, and it sounds a lot like Zorba the Greek.

Haddock and cod, along with other species of groundfish that live near the sea bottom, both use their swim bladder as an internal drum, producing a distinctive *knock, knock, knock* sound. In haddock, the more sexually aroused a male gets, the faster his internal drumming. When the knocks come so quickly they form a continuous hum, the female knows he's ready to pop. The question is, is this the guy she wants to spawn with?

To convince her the answer is yes, a male haddock adds some smooth moves to his repertoire of beats, providing a bit of eye candy to go along with the primal thud of his swim bladder. When mating season rolls around, males gather on spawning grounds in enormous numbers, sending forth a percussive call that may work to lure females in from the murky waters. For fishers, the aggregations offer a prime opportunity to harvest this popular seafood, a major contributor to the endless plates of fish and chips in Britain, fish sticks or "scrod" in the States, and especially loved in Norwegian "fishballs."

Though more research is needed, it is likely that, similar to peacocks and some mammals, cod and haddock may be lekkers. It's not as dirty as it sounds. A lek is a group of males that get together in order to attract the attention of females for mating purposes. Picture the pick-up beach volleyball scene in early summer: groups of guys strip down and show off their muscles, providing the ladies a convenient and conscripted location to observe the competition.

Leks form in species where males don't really have a way of guarding or isolating females or securing a resource that would attract a female. For male haddock and cod, the sandy plains hundreds of feet below the surface of the Atlantic don't offer much to work with, so they group together to draw the females in and then compete for affection.

As these fish live in deep, dark waters, nobody has managed to directly observe their mating behavior in the wild. But researchers have witnessed their seduction sequences in the lab. In haddock, a male begins by "patrolling," swimming in tight circles or figure eights just off the bottom as he emits a slow and steady drumbeat. The constant circling allows bigger, more dominant males to establish territories within otherwise nondescript mating areas.

A female approaches, hovering over the male or dropping down to his level, eyeing his maneuvers. She then swims off, the male following her closely. But he can't show off his goods from behind, so he swims alongside her and flicks his vertical fins up and down for a few moments; then he crosses in front of her so she can get a good look at his flank. As he progresses with these moves, circling the female, flaunting his stuff, he sends out those distinctive knocks faster and faster.

Sometimes, the male returns to swimming his figure eights over the bottom, and the female comes back for another look. Eventually, minutes, hours, or even a day later, the male moves in for his final number. Swimming upside-down underneath her, the male orients so the two are belly to belly. He clasps his small pelvic fins (located up by his chin) with hers, and holds her close. His pulsing drumbeat has now crescendoed to a mighty hum, the vibrations perhaps helping to synchronize the final act.

Pointing their noses upward to hang vertically in the water column, the two lovers thrash their tails and press their vents closely together. Then they spawn. The male's extensive displays, ever-increasing speed of drumming, and a final pelvic clasp all may help the pair coordinate this simultaneous release.

Cod sex also culminates with belly-to-belly bursts of millions of eggs and sperm (haddock and cod are rather prolific), but in contrast to haddock, studies on cod reveal the males wait until the female arrives before they begin their show. A female cod takes her seat by lying motionless on the bottom. It's a move that says, "OK, boys, you may dance for me now," and dance the males do, circling overhead as she watches from below. Having established ranks through aggressive chasing, prodding, and some biting, dominant male cod manage to circle waiting females more often than subordinates. After the displays, females usually swim off, but within about twenty-four hours, the performing male will find and attempt to mount her. Swimming alongside her, the courting male performs a series of jerky dashes out in front of her, accompanied by some lusty grunts before initiating the vertical mount.

For both female cod and haddock, the ability to size up a mate in close proximity may help females assess not only male fitness, but also size and rank. Given the preferred sexual position—a sort of vertical missionary style—having a mate equal in size helps to align the openings where sperm and eggs pour forth.

And here's where the unexpected impact of different fishing techniques can cramp a cod's (or haddock's) seduction style. In both species, males and females tend to hang out at different depths. Male cod show off their moves up in the water column to females watching from below; haddock males get their groove on along the seafloor and wait for hovering females to swim down and join the fun. This means that for both species, fishing gear that targets a certain depth—say, bottom gear such as trawl nets or midwater gear such as hook and line—will capture more of one sex than the other, skewing the sex ratio of the population. That in and of itself can make it harder for the remaining sex to find mates and reproduce.

In species where males perform to impress a female, the simple act of displaying may also expose these macho males to more risk. All that dashing and flashing, all of the charging and grunting at intruders coming into their performance space—all of these behaviors likely lead these fish to hit baited hooks or swim into nets more readily than more timid males in the population. The result is similar to what happens in the aquarium trade with mandarinfish, where flashy males draw the attention of divers who preferentially select them for their bold colors and size: females are left with scrawnier suitors to choose from.

Like growing up in a small town where most of the available singles are likely some form of close cousin, many marine species face slim pickings for attractive mates these days as a result of the way we fish. When we prefer—or our fishing gear selects for—the sexiest or most dominant individuals in a population, we wind up leaving behind a population of weaker, less attractive options. And, no surprise, sexual activity and productivity take a dive. Fish, it seems, like most other species on the planet, would rather reabsorb their own gametes than waste them on an unworthy mate (try that line the next time you need to firmly reject someone hitting on you at the bar). With the loss of the sexiest males, a female may need longer to find a suitable partner; she may be forced to compromise her standards and settle for spawning with an inferior mate; or she may skip spawning all together. In all of these circumstances, less sex or less successful sex happens as a result.

There are ways to manage for these effects, but doing so requires knowing that these behaviors even exist. For many species, this kind of detail is missing—researchers either don't yet know the mating rituals of the species or haven't built these conditions into their models for managing fishing impacts.

Meanwhile, it would be a shame to imply that the only love songs sung in the sea were made by grunting groundfish. Courtship for humpback whales, for example, contains heartfelt singing of rather complex tunes combined with intensive battling, sometimes in teams, as males race to successfully score with a female. If the long-distance love songs of blue whales evolved to advertise to mates from afar, the songs of humpbacks are more like intimate serenades: sung on the breeding grounds where females congregate to give birth. The only thing is, it's not clear who exactly these males are singing to.

Traveling thousands of miles from their feeding grounds in the colder climes (where rarely a song is sung), Hawaii's humpbacks migrate in the fall to winter breeding grounds in the tropics. As more and more males arrive in the warm waters, a chorus, sung by a male-only choir, erupts. These songs are sung so frequently, and travel so well through the turquoise waters, that anyone on the leeward side of Maui or Big Island can walk out to just past the surf break on any beach, dunk their head underwater, and hear the haunting tune.

Males sing songs of multiple themes that can last from five to more than twenty minutes, and at any one time, all males in a population sing the same song. During the course of a breeding season, songs often evolve, sometimes quite quickly. It's common for new themes to take over an older version, like a mash-up of a Taylor Swift number with a Sinatra standard. And just like on land, humpback whales also produce hit singles that sweep across the sea. For some reason—perhaps due to the larger size of the populations—completely new songs tend to move from west to east across the

Pacific, with humpbacks off Hawaii picking up on a song years after it was first belted out by a group of males off, say, Fiji.

We still don't know why the males sing (although satirist Christopher Moore in his book *Fluke* offers up an amusing hypothesis that also solves the mystery of Amelia Earhart's disappearance). Most of the time, the singer is alone, crooning the ethereal song into the sea. And most of the time, the only thing he seems to attract is other males. Sometimes a singing humpback is joined by another male and the two may stick together, even cooperate to chase down a female. Other times, the singer may swim over and join with a group of males who then further merge with a bigger party of males to then go pursue a female.

During these large heat runs—an all-out race where males jockey for position nearest a bolting female—there is far less wooing, and a whole lot more shoving and pushing going on. Pectoral fins out like elbows, the males knock their way forward in the crowd, slicing with their tails, ramming with their heads. There's no singing now. With males each the size of a large bus and weighing about forty tons, these are some of the biggest dating matches on the planet. Who wins this race, and how that winner is determined, remains a mystery.

This is surprising, given how many voyeurs now head out to sea to watch the whales cavort on their breeding grounds. Despite the ever-increasing tourism trade that puts more and more people front and center to their dating games, despite the fact that they aggregate on specific breeding grounds, and despite the fact that the waters are crystal clear . . . nobody has seen two humpback whales actually in the act. (If *you* have, please do tell!) Like the vast majority of all marine species, we still don't understand the process by which a male succeeds in those final steps of seduction.

Even those species we've studied well continue to surprise us: it took hundreds of underwater hours before Warner and colleagues noticed the usurpation of nests and then total abandonment by those duplicitous wrasse. Fish expert Hastings discovered the neat-freak nature of tube blennies only after years of observation. For most of these species, we simply don't have that much time logged yet. Thus we can't always know how our actions may be affecting the outcome of their pursuits, sexual or otherwise.

In the case of humpback whales, however, the results on our most recent impacts are in, and they are looking good. Since the international ban on whaling went into effect, a majority of humpback whale populations worldwide are increasing in number, proving we can be an uplifting force for some of the biggest sex in the sea.

SNEAKING SEX

Seduction can be overrated.

Some males and females skip the costly and complex courtship and simply steal, sneak, or slip in some sex on the sly. The masters of such deception are cuttlefish, who have proven themselves to be truly talented transvestites.

Cousins of squid and octopus, cuttlefish are cephalopods, capable of near-instantaneous changes in both color and body shape, including the texture of their skin. Some small and crafty males put these powers to use in order to sneak sex with a female right under the arms of her big, macho boyfriend. It's an effective strategy in the battle for a very limited pool of mates.

On the shallow breeding grounds of Australia's southern coast, giant cuttlefish gather by the tens of thousands to mate during a short four- to six-week period. Despite the enormous numbers, it's a challenge for a bloke cuttlefish to find a lone female to partner with—there are about four males for every female on average and sometimes as many as ten to one. In addition to these tough odds, females can be rather finicky, rejecting up to 70 percent of all males that approach, with no apparent rhyme or reason behind their selection criteria. For a small male cuttlefish, the odds of finding and fertilizing a female are even slimmer as most females are already paired with much bigger consort males. That is, the odds are slim for a small *male* trying to outcompete big males. As a "female," however, they score a fair amount of fertilizations, which means there's a whole lot of cross-dressing male cuttlefish on the reef.

Besides their smaller size, female cuttlefish can be distinguished by their mottled brown coloration and shorter arms—features easily mimicked by a small male. Swimming calmly nearby, a cross-dressing male need only don these subtle female tones and tuck in his arms—taking special care to hide his modified fourth arm, the one all males use to hand sperm packs over to the female—to easily fool a bigger male into thinking he is a she. By assuming this disguise and acting like an egg-laying female (who repels all mating attempts), these con artists can sidle right up to a female despite the presence of her consort nearby.

As soon as the large male is distracted by a challenger, the small male moves in. He immediately shape-shifts, ditching his feminine garb in exchange for courtship colors, a sexy series of cascading stripes and waves rippling across his body. If all goes well, the female will allow him to mate with her. She often does, perhaps impressed by his cunning ploy. Being small can have its perks for a male.

If they are caught in the act, the intruder male may suffer a good beatdown by the bigger male. But if he is quick enough, the sneaker male will flash back on the female disguise and successfully avoid the fight. Such deception may lead to another problem, though, as the small cross-dressed male trades a skirmish for a lustful lunge: many female mimics are so convincing, big males try to mate with them.

In mourning cuttlefish (named for dark circles that often rim their eyes), the deception is taken a half step further.

A male seduces a female by flashing flamboyant zebra stripes along his flank and back, a signal that he is horny and hopeful. Such displays, however, catch the eye of other nearby males who may then dart over and intercept the interested female and mate with her first. So if another male is present, a courting male will instead place himself between the female and the other male and then doubly display: on the side of his body that's facing the other male, he presents female markings; on his female-facing flank, he broadcasts a suave, sexy "I'm all male, baby" display. Viewed from above, the cuttlefish's back is split straight down the middle, with jungle safari zebra stripes (always sexy) on one side and mottled brown and unassuming female pattern on the other. This semi-cross-dressing male thus courts his lady in plain sight of another male, without raising any suspicion. When more than one male arrives on the scene, however, the masquerading male drops the act and reveals his true colors—the ruse only works from a narrow perspective, after all.

Cross-dressing not only aids males looking to sneak in some sex on the sly. It also works for females trying to avoid pestering males. And when trying to look like a guy, nothing is more convincing than donning a pair of testes.

In at least one species of squid, the females fake having male parts by displaying a pattern on their body that looks like the shading created by the elongate testes of males. They essentially superimpose having male parts on the inside by changing the color of their skin on the outside. And it works. Females with testes-like patterns don't receive the same level of harassment by courting males.

All of this masquerading works to help small males score some otherwise unattainable action (as well as assist females with avoiding it). But not all small males rely on such clever deception. Instead of outsmarting big dominant males by temporarily donning female coloration, some male fish go for a much more blunt approach: they simply out-sperm them.

Since the vast majority of female fish release eggs into the water (or an open nest), nearby males can streak in to add their sperm to the mix. Of course, it's a gutsy move to storm in on a big alpha male just as he's about to seal the deal with a female. These streakers must be fast, nimble, and have some serious cojones. No, really. Their success depends on developing absolutely enormous testes.

In many species, two male reproductive roads diverge in the sea: one path puts energy into growing big and tough in order to win near-exclusive rights to some female affection; the other stresses staying small and swift and just growing big male parts. The cigar-shaped bluehead wrasse, a common species among Caribbean reefs, has both alternative male types: a big supermale, also known as a terminal male, or TP, and "initial phase" males, or IPs.

There are a lot of strange twists to how a male becomes a TP male or stays an IP male. Suffice to say the supermales are brassy males with bold blue heads that defend prime locations on a reef where females prefer to go for their daily lunchtime quickie. The TPs are much bigger than the IP males and far more aggressive. IP males lack the namesake blue head, their coloration often resembling that of juveniles and females, as does their body size. Instead of growing big bodies, IP males invest heavily in growing their reproductive parts, with a whopping 20 percent of their body weight composed of testes.

If men were proportioned like IP bluehead wrasse, a two-hundred-pound guy would have to fit forty pounds of testes into his pants. Good luck with that.

These enormous sacks allow an IP male to jettison up to fifty million sperm in one go—supermales only release three to four million, which makes them seem, perhaps, a bit less super. (Should an IP male grow big enough to start winning some battles, they switch to become TP males, at which point their testes shrink way down).

At the end of the day, the two tactics work to secure fertilizations with a female under very different circumstances. Arriving slightly ahead of the females each day, a supermale first aggressively chases away all other males from a favored spot. Then he welcomes the females with vigorous vibrations of his pectoral fins and a few dashes up in the water column. A female impressed with both his location and his moves will join in one of these upward surges. The pair will swim almost belly to belly and simultaneously release their sperm and eggs into the upper waters. Successful TP males have been known to exceed 150 matings in a single day—which is even more impressive given they don't spawn all twenty-four hours long—these fish tend to be noontime lovers.

In order to maximize his sexual output, a TP male not only defends a prime territory and seductively woos a female with upward thrusts in the water column, but he also does a little math to ensure he doesn't run out of sperm. On reefs with only a few females, a TP male will let loose lots of milt (the term for male fish semen), perhaps four million sperm per female if there are about twenty females around. This results in about 98 to 99 percent fertilization success. But should the number of females double, as was the case thanks to another set of experiments by Warner, the TP male will adjust. The first day that the additional females arrive, the TP male spawns as usual and, alas, runs out of sperm by somewhere around the twenty-fifth female. He'll continue to perform the spawning rush with the rest of the females, but his tank is dry. However, the very next day, knowing the larger size of this new batch of daily visitors, he will adjust his sperm contribution for each female down a notch. With fewer sperm released, fertilization success per spawning drops to about 92 to 93 percent, but by being able to fertilize twice as many females, the TP males' reproductive output overall goes way up.

A nearby IP male that notices all the commotion happening around a TP's territory may be tempted to dart over to streak through the spawning couple's bliss and add some of his own. But a courting supermale is not a fish to be trifled with; he will break off his ascent with a female to lunge at any intruder who may be trying to get in on the action. So the IPs have to wait for these supermales to be past the point of no return, so to speak. They must dash in and release their own DNA into the mix right at the apex of the pair's vertical rush, just as they too release clouds of sperm and eggs. This strategic streaking works well enough, but only results in about 50 percent of the fertilizations, despite the male dishing out ten times as much sperm. Proper positioning makes a big difference, and the TP male takes the advantage here.

On some reefs, there are simply too many wrasse and not enough prime spawning territories. Under these conditions, gangs of IPs will overrun supermales and chase down females directly. This can result in one female spawning with several, even up to a dozen males, at once. In this lottery-type circumstance, with multiple males puffing out enormous clouds of sperm, the more sperm a male can dump out, the greater the chance some of his DNA will make it into the next generation. Small IP males with prolific sperm counts thus win big on these more bustling reefs.

Such sneaky males exist in more than pretty reef fish; they also ensure the future of some of the world's most popular seafood. Both coho and chinook salmon have sneaker males, known as jacks, that ditch out on life at sea early (or altogether) to return to their freshwater environments after only a few months. There, these diminutive devils wait to ambush pairs of larger spawning adults, spewing their own milky milt into the streams. Too small to attract females themselves, they time their faster-than-average ejaculation to occur right when the couple climaxes, and then scram before the enormous males can retaliate. Dastardly as this behavior appears, jacks are likely critical to the long-term survival of salmon. These precocious little males, because they are younger than the spawning couple and thus from a different generation of parents, likely help increase the genetic diversity of the population by adding some of their sperm to the mix.

SECURING A MATE through scent, sound, sexy shimmy, or on the sly are all part of the seductive repertoire beneath the waves. But there is one more strategy of undersea inhabitants that plays a critical role in securing sexual opportunity: swapping sex. By this, I do not mean some kind of underwater swingers party; instead, individuals in these species have the ability to change sex during the course of their own lifetime. The pathway to sexual success involves more than just seduction; it requires some critical calculation to determine not only who to seduce, but also when—and as which sex.

3

FLEX YOUR SEX

Sex Change in the Sea

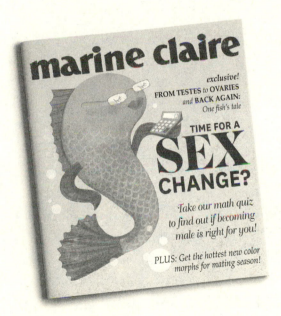

SEX-SEA TRIVIA

- *Some grouper can start life as female and then morph to male. Some oysters can do the opposite.*
- *Clownfish put an* Oedipus *twist on* Finding Nemo.

- *Sex change via peer pressure is common in the sea.*
- *You are what you eat: in one shrimp species, becoming male or female depends on how much seaweed you munch.*

SEX-SEA SOUNDTRACK

1. "Take a Walk on the Wild Side"—Lou Reed
2. "Dude (Looks Like a Lady)"—Aerosmith
3. "Eye of the Tiger"—Survivor

Once Upon a Time there was a King and Queen who ruled over a peaceful kingdom. The peace came from order, and that order was imposed through fierce intimidation. No one dared rise up to challenge their reign. Beautiful and standing a full foot taller than the King, it was no secret that the Queen was in command. It was even rumored that she bullied her King, just as he bullied his court.

Then, one night, the Queen died. Within moments, a strange new force began to swirl within the castle walls. As if released from some spell, the King felt a change deep within himself, a blossoming of something new, something different, something . . . feminine. For the next few weeks this inner transformation progressed until finally he stepped forth as a new and powerful Queen, equally as beautiful, fertile, and commanding as the former had been.

The King-now-Queen took as her mate a spirited youth who had, under the same spell, developed into a strapping, virile male. Under the new Queen's hard-hearted watch, the new King embraced his role and began a new reign of intimidation—and the two lived and laid many successful clutches of eggs, Happily Ever After.

The End.

OR SO A FAIRY TALE might go, had the Brothers Grimm known anything about clownfish. Yes, clownfish. Sorry folks, but Pixar got it wrong. Way wrong.

When it comes to the relationship dynamics of clownfish, the true adventure tale reads more like the Greek tragedy *Oedipus* than it does *Finding Nemo.* As Nature writes the story, by the time Nemo hatched out of the egg, his dad, Marlin, an unpartnered male head of household, would have already morphed into Marlene. For clownfish, when the leading lady dies, the top dog promotes to bitch.

Rather than chasing after a kidnapped Nemo, Marlin-now-Marlene would stay at home and welcome the next largest male around to join her as her chosen mate inside her spacious anemone abode. A mature, ready, and waiting female occupying a decent anemone would not remain lonely. Nemo, if he ever did escape and make it back home, would find the anemone filled with other male clownfish. He would have to wait his turn to meet (and mate with) his father-turned-mother, delaying the happy reunion of son-as-lover with mother-who-was-father.

Though lacking the sharks, jellyfish forest, and surfing sea turtles, when it comes to personal growth and triumph sagas, the real Nemo story offers a far more colorful tale that hinges on the ability of a clownfish to change sex during its lifetime. It's a strategy deployed by many species of fish and invertebrates—species that never have to wonder what shagging is like for the opposite sex . . . they know.

A BRIEF SOJOURN INTO
SEX-CHANGE STRATEGIES

Under the sea, the boundary between male and female is far more fluid than on land. A little midlife sex swapping is part of the natural life cycle of everyone from Nemo to the shrimp in your shrimp cocktail. In fact, start naming all the sex-changing animals in the sea, and the list looks like a recipe for bouillabaisse: mussels, clams, shrimp, and a whole slew of fish. There are others, too, such as worms and some sea stars (formerly known as starfish), that don't lend themselves as readily to a chowder but do exhibit some serious flex in their sex.

Though energy intensive, the ability to alter one's sex is a strategic way to boost babies per reproductive bonk. Here's why: in some situations, one sex will make more babies when they are bigger (or older) than when smaller. In human terms, a woman's fertility peaks in her twenties and declines later in life. But the same doesn't go for a guy. Instead, he can continue to make babies by hooking up with younger women far into his fifties, sixties, and beyond.

Now imagine that younger guys, with their overeager sex drive and lack of experience, weren't likely to get much action from discerning females who wished to have their babes sired by only the strongest, wisest, best providers. Under these circumstances, if people wanted to maximize the number of children they could produce, and if we *could* change sex, it would make sense to start life as a female, making babies by reproducing with older men while you are young and fit. Then, as conception and baby-carrying success diminished around thirty-five years of age, you would switch to being a male, and kick up your offspring output by finding a pretty young thing to mate with. Voila! You've just increased your human production potential.

Of course, you also would have to endure the pangs of puberty *twice*. In reality, human biology is too prudishly rigid to allow for this kind of flexibility with our sex. The same holds for other mammals, including species such as elephant seals, which would certainly benefit from that kind of flexible strategy: all those small males kicked off the beach by the big alphas could instead start off as females and then morph to males when big and ready to do battle. Alas, it is not an option for most vertebrates.

Fish are an exception. Along with many invertebrates, they aren't nearly so limited. For them, the cost of sex change is a small price to pay in return for some serious reproductive advantage.

THE INS AND OUTS OF
HOW SEX CHANGE WORKS

Technically speaking, all sex changing species are considered hermaphrodites; but instead of *simultaneous* hermaphrodites, where one individual has genitalia and gonads (the internal sex organs) for both sexes at the same time, sex changers are *sequential* hermaphrodites, maturing first as one sex and then changing into the other. Different factors of a species' ecology and social system dictate whether changing sex makes sense, and in which direction a species is likely to go. Oh, yes. Sex change goes both ways beneath the sea.

Animals that start as males and turn to females, such as the sex-tower-building slipper shells, are *protandrous* hermaphrodites. "Proto" means first, and "andro" means male, in contrast to protogynous hermaphrodites, which start life as females and turn into males, with "gyno" indicating female. And in a few species, individuals are able to flip-flop back and forth between the two sexes their whole lives through. A little sequential sex-change poem to help it all sink in:

Andro *from the ancient Greek*
Means anything that's male.
Proto's *the preface that you'll find*
Marking the start of a tale;
So Protandrous *folks are those*
Who begin life as a guy
And later morph to female
(There are many reasons why).
But some species start out first
Embracing female persuasions;
These Protogynous *hermaphrodites*
Turn male later, given certain occasions.
Some may rely on a partner
To tell them when to switch,

Or sometimes it's the neighbors
Who give individuals the itch.
From hormonal cues to harem size
There are many factors at play;
Different species use these clues
To guide them along the way.
These sex changers have the answer
To our own popular mind-bender:
What is sex really like
For the opposite gender?

Much of what determines the lifelong sexual strategy of an individual relates to how resources and mates may be monopolized within a population: Are there loners that group up for open orgies or do a few males dominate a whole bunch of females? Are the odds of finding many mates high, or is it best to just settle down with the first one you find? Depending on the answer to these and other questions, species will employ a range of strategies for firming up their sexual prowess.

To be a true sequential hermaphrodite means 100 percent conversion—in body and behavior, inside and out. This means all your parts—including the private ones—would indeed change. Fish and invertebrates undergo an extreme makeover, dramatically reconfiguring external packaging—in some fish, males have genitalia that are long and dangly, too—internal organs, color, body shape, and behavior.

The queen parrotfish is a colorful example of such external change and a fish many people are far more intimately connected with than they realize. Parrotfish are the ocean's equivalent of a weed whacker. Unsung defenders of coral reefs, they use their beak-like mouths to bite off clumps of seaweed and thereby keep algae from overgrowing the corals. In the process, they inevitably bite off some coral, too, and their guts grind it into powder. Like an airplane leaving contrails, parrotfish release white puffs of coral dust as they swim along their way. Eventually, the pulverized coral remains are washed up on the coast, forming soft seashores. If you've ever sat on a tropical white sandy beach, you've lounged upon piles of parrotfish poo.

As a young female, the queen parrotfish displays earthy, subtle tones that help her melt into the shadows of her coral-walled home. But after a few years, she trades in her dull dress for a carnival of colors that would make Sir Elton John swoon. Mauves and tans are bowled over by turquoise, radiant blue, and bright yellow as she sheds her feminine mystique to erupt into full-blown masculinity.

Sinewy, snakelike ribbon eels offer similarly impressive color transitions to accompany their sex change in the opposite direction. Jet-black juveniles with a prominent yellow racing stripe down their backs mature into bright blue males. The yellow stripe remains, affording them the look of a Swedish flag. A few years later, the bright yellow stripe widens, eventually enveloping their entire body as the eel morphs into a fully functioning and golden female.

In some species, the transformation goes beyond a simple costume change to include an element of shape shifting. One of the world's largest coral reef fishes, the Maori wrasse (*Cheilinus undulatus*) also has the nickname "humphead." Though the name does refer to a bony protrusion, it's not what you think. Instead, the oldest and biggest males—some may exceed seven feet in length—have a brilliant blue brow that juts out prominently, casting their eyes in perpetual shadow. But most didn't start out that way. Many of these Neanderthal-like foreheads blossomed from the smooth, streamlined red and orange domes of their former female selves. Sex change in this species brings not only new colors, but also a crested cranium.

In all cases of sex change, the extreme external rebirth reflects an internal shift equally as impressive, for it involves the complete reconfiguration of all sexual organs. In general, the new "machinery" will develop first and become functional before the old gear is reabsorbed. So in protogynous hermaphrodites, sperm-producing testes develop to replace the ovaries, which then start to dissolve. For protandrous hermaphrodites, the opposite occurs, as new ovaries grow and the body begins to reabsorb the testes. Behind this physiological tango between gonads past and present dance a chorus line of hormones. These trigger different steps in the process of transition. What triggers these triggers depends on the species.

LI'L MALES AND BIG OL' FAT FECUND FEMALE FISH: PROTANDROUS HERMAPHRODITES

To understand why the real Nemo story reads more like Greek legend than it does Pixar, first you need to wrap your head around this fact: the bigger a female fish grows, the more eggs she can make. This "bigger equals more eggs" concept is completely foreign to humans. Our females are born with roughly the same number of eggs—about one million. No matter her height, weight, ethnicity, et cetera, a woman has all the eggs she will ever have *before* she is born—by about twelve weeks in utero. As she ages, the number of eggs she carries goes down. By the time she hits puberty, only about half of her lifetime egg pool remains (most of these never fully develop and are reabsorbed into the body—only about three hundred to five hundred eggs ever fully mature).

This is not the case in fish (and many other marine species too). For them, eggs are produced over the course of their female lives, and size matters. The bigger a female is, the more eggs she can fit, and as long as she is healthy, the more eggs she can make. For example, a fourteen-inch-long vermilion snapper will make about 150,000 eggs. A twenty-two-inch female of the same species will make 1.7 million eggs. That's over ten times more eggs before the fish doubles in size. So big, old, fat, fecund female fish (affectionately known as "BOFFFFs") can pump out far more eggs than their younger sisters a few notches down the size scale.

There may be other advantages to BOFFFFs besides sheer increases in number. Older (and wiser?) females may release eggs over a longer spawning period and more diverse spawning habitats than younger females. This helps them hedge their bets in terms of hitting favorable conditions for larval survival. These factors indicate that BOFFFFs are disproportionately beneficial and perhaps even critical to long-term survival of populations.

That bigger females can carry more eggs is not a trick specific to sex changers. Any large female fish—whether she was born female and stayed that way or started as a male and transitioned to female—has the potential to contribute significantly more to future generations than a smaller female. But this feature becomes something sex-changing species can exploit, especially for those species that engage in the abnormal behavior of pairing up one-on-one for the season. Such is the case with clownfish, male-to-female sex changers that join seahorses in the minority club of species forming monogamous couples.

As a candy-colored bite-sized fish on a predator-filled reef, clownfish (also known as anemonefish) tend to stick within the confines and safety of their anemone homes. Distant cousins of jellyfish, anemones have a soft body surrounded by rings of stinging tentacles that present a perfect fortress of protection for clownfish, which hide within their waving tendrils. But a good anemone can be hard to find. If you're an adult clownfish who decides to go looking for a new home, other clownfish will likely chase you away from their already-occupied abodes. No room at the inn for you. But as a juvenile, you're pretty innocent and pose no competition to the ruling adults, so unless an anemone is particularly crowded, odds are you can stick around. Thus, young clownfish use their sense of smell to find their way to a good anemone, and once allowed to join a group, they stay.

Confined to an anemone, these fish are stuck with whoever else lives there. It's kind of like being forced to date only the girl or boy next door. But although there may be four to six individuals living around one anemone, only the two largest individuals will mate: the one and only female with the largest male. And here's where being a BOFFFF comes in handy. Generally speaking, even a small male has enough sperm to fertilize all of a female's eggs. The more eggs the female can make, then, the more offspring the couple can produce. So a bigger female benefits them both. By starting off as a male, an individual that hooks up with a big female can produce lots of offspring when small, and then, when his older, bigger partner dies, he can then grow into the female role, get a new mate, and continue the high-level offspring output. This is what clownfish do, and this is why the real Nemo tale doesn't look anything like the movie.

The trick to an individual clownfish's sexual success, though, isn't just sex *change,* it's also preventing other adult clownfish from sneaking sex with their mate. Both the top-ranking male and the female engage in some psychological warfare, bullying the other resident clownfish and stressing them out so much that their sexual development ceases. It's a delicate art form, really. The female torments the largest male just enough to keep him from growing too big (and risk turning into a competing female) but does not intimidate him too much, which would suppress his manliness. Whipped as he is by the female, the male then takes out his aggression on the next largest individual, but he goes all the way, intimidating that male into suspended maturation. The intimidated then becomes the intimidator, turning to dominate the next biggest, and so on down the line, ensuring that each individual knows his place in the pecking order and remains in pre-pubescent limbo.

Life may not be easy for young, stunted clownfish, but there are advantages to all that torment. When the female dies, the ranking male can quickly convert to female and reap the reproductive rewards: the next juvenile in line simply rises up, grows a pair, and the new couple gets on with the show. Nobody has to venture outside the green zone of the anemone to find a mate.

Clownfish are not alone in their protandrous lifestyle; many oysters also know sex—in the biblical sense—from both sides of the bed. The most popularly consumed species, including those Bluepoints and Belons, Sweetwaters and Wellfleets, Kumamotos and Pemaquids, in all their wondrous, buttery, salty, smoky, earthy, fruity "merroir"—all have the potential to morph from male to female. Such a talent is also beneficial when you're an animal that's stuck in the muck for life.

Glued together as a living rock wall, oyster reefs or "beds" are made up of generations of individual oysters that, as tiny free-swimming larvae, sink down from the surface to attach to and grow on the backs of their ancestors. During a season of summer lovin', oysters contract the two halves of their shelled house, forcefully ejecting enormous numbers of sperm or eggs into the water, where they mix with the gametes of other neighboring oysters. As we will discuss in a few chapters, animals such as oysters that can't move instead set their gametes free in the open blue to find their complement and fuse. To help increase the odds that fertilization will occur, these animals pump out spectacular numbers of eggs and sperm. Bigger females are advantageous because they can significantly up their egg output—just like BOFFFFs. An adult female oyster may release over a million eggs in one go, and they often have multiple spawning events in a year. Smaller males, with fewer energy reserves, can still make lots of cheap sperm, but it is difficult to make lots of fat-rich eggs and still have energy left over to grow. So, protandry makes sense, with bigger oysters able to divert more energy to female reproduction, which helps everyone.

While size does matter, it is not the only factor controlling sex change. Social cues are also important. As Dr. Juliana Harding, an oyster expert at Coastal Carolina University, notes, "What's the point of spawning as a male if everyone around you is a male?" Or, equally important: why bother changing sex if your neighbor already has? Harding explains that oysters use chemical cues to determine who else is around and of what sex in order to calculate when and if sex change makes sense. "Both size and social cues influence the end product."

Actual sex change happens after spawning, when the overflowing gonads are finally spent. But in contrast to many sex changers that can make the leap from male to female (or vice versa) in a manner of days, oysters take a bit more time in swapping sex. Harding explains: "It is like phasing out one set of equipment and bringing a new set on line and testing it before getting rid of the old. From an evolutionary standpoint, you always want to be able to spawn as something so as to not miss opportunities to contribute to future generations." This is also why oysters "trickle spawn," releasing some eggs or sperm over an extended time frame.

Not all individual oysters change sex, though. In some species, oysters born male stay male while others born male will transition to female after a few years. The difference in the two paths seems to be mostly genetic, with some environmental influence. In the Pacific oyster, for example, it is likely that males born with a genotype MF are true males (similar to the male human XY genotype); those oysters with an FF combination (similar to the female human XX genotype), however, are protandrous—born male, they may sex-change into female after one or two years. There isn't a hard deadline for when sex change occurs. Instead, both age and the environment can influence the timing: if food supplies are low, or the conditions otherwise harsh, an individual may delay the switch. On the other hand, disease or fishing pressure that targets older and larger individuals may trigger an earlier transition to female—to ensure they pump out a few rounds of eggs before being knocked off. This kind of variable sex change is one example of how external forces can fundamentally affect the sex lives of a species.

The male-to-female transformation is but the tip of the sex-change iceberg, however. The far more dominant strategy (at least in marine fish) for boosting sexual success via sex swapping is the female-to-male route— that's the pathway of choice for species where big males can effectively rule the school.

HAREM MASTERS AND REAL ESTATE MOGULS: PROTOGYNOUS HERMAPHRODITES

For species where male domination of females or a territory is key to sexual success, bigger is better. Without the brawn, it's nearly impossible to successfully defend a group of females or a prime mating spot that would attract the females to you. In these kinds of mating systems, it pays to first mature as a female when small, mate with the dominant male for a few years, and then, when you've gained enough girth, and experience, switch sex to become a strapping, imposing alpha male and start mating with all the females. This is the strategy behind protogynous hermaphrodites, species that start off life as females and then transition into males. It's a switch that requires careful calculation. After all, taking on a dominant male role can be costly. First, she has to learn how to be an effective male from a behavioral (and attitude-wise?) perspective—both to compete with other males and to win over females. Then she has to expend energy building new male reproductive parts and dismantling the old female structures. So how to know when it's time to take the plunge?

For some species, it all comes down to owning killer real estate.

Males that can successfully guard prime territories have the females come to them. We saw this in those males that seduced females with sweet spots for laying eggs. Such select locales can also be places overflowing with food or containing the perfect conditions for free-spawning eggs into the sea. It's the equivalent of having that family-friendly home in a good neighborhood.

The cleaner wrasse provides one example. About the length of your hand and the typical cigar-shape of a wrasse, these fish have a yellowish head, bluish rear, and a prominent horizontal black stripe that runs from eye to tail, thickening at the end. As their common name implies, these coral reef fish stick around discrete "cleaning stations." These hangouts are like the local salons of the reef, only instead of getting a haircut or a wax, it's the spot to go to get yourself debugged. Fish wanting such a service need only show up and assume the proper position, hovering with fins outstretched and mouth agape. The wrasse will then go to work, picking little parasites or dead and damaged scales off the body of the visiting fish. These cleaning stations are such neutral ground that the tiny wrasses will even swim in and out of the open jaws of large predators, such as grouper. And just like the neighborhood beauty parlor, the customers are loyal— often revisiting a station on a daily basis.

So, if you are a male cleaner wrasse that can defend a group of females around one of the hottest salons on the reef, you are in business. Female wrasse will stick around servicing you along with the clients. In these wrasse, the lasses are loyal. They remain part of the harem even if the male bites the bullet. The dominant female simply starts playing the male part—within an *hour* of his departure.

I said they were loyal, not sentimental.

Just imagine life as a male wrasse for a moment. Sure, there is lots of lovin' to be had as harem master. But want to get away for a holiday? Best to take the gals with you. Otherwise, you're likely to come home to find out your formerly submissive dame has pulled a Don Juan and seduced all your females into her—now his—own harem.

Nature does give the guys a small break: should the male slip out for only a few hours, the largest female may get a bit frisky with the other gals, but she will resume her ladylike posture promptly, no questions asked, upon the male's return.

For a female that must replace a male, her first priority is keeping the harem together in order to secure future mates. Because it can take two to three weeks for her to grow functional male parts, acting like a male may be the best option for ensuring that the other females keep producing eggs. That way, when the sex change is complete, she-now-he can finally get down to business. The ploy works. Smaller females just don't seem that discerning. Their instinct is to mate with the largest, dominant individual in the group. As long as she acts like a he, this seems to be enough.

The opportunity to dominate a desirable undersea domain is one reason a female may transition to male. But for several species, sex change is not an individual's choice; often, it's triggered by the number of other females and males hanging around. Think about that one for a second. It would be as if, following a move to a new apartment, your sex would hinge on how many women and men lived down the hall. Not only does this system leave your sex (and sex life) dependent on the vagaries of larger social dynamics, it also means you need to pay close attention to the numbers.

SEX CHANGE CALCULUS

Some fish and shrimp have to be good at math; their sex lives depend on it.

Take the popular aquarium fish known as a sea goldie, or lyretail anthias, a small coral reef fish that can keep track of quite a large number of their schoolmates' size and their sex. Remove a few dozen males from a reef and nearly the exact same number of females will change sex to replace the lost males. What makes this practically perfect conversion ratio even more impressive is that it can happen across dozens of individuals and in an orderly fashion, with the largest females changing sex first, and following females converting according to their rank, by size, right on down the line. That's some pretty impressive calculating.

The bucktooth parrotfish, however, takes the math skills a step further. A protogynous sex changer, the biggest female would usually change sex to take over the harem and reap reproductive rewards whenever the residing male leaves the scene. But this doesn't always happen. Often, the largest female chooses to remain a dame. Such a decision results not from lack of ambition—quite the opposite. It seems to depend on a deliberate calculation of the relative size of other females in the group—and how many eggs they may be churning out—as well as the degree of male competition.

As a newly transformed male, the biggest female would exchange the total number of eggs she produced in order to fertilize the eggs from all the females in the harem combined. You might assume that the total number of eggs of all the other females in the harem surely would be more than the number of eggs of the single biggest female—but then you'd be underestimating the power of the BOFFFF. It is quite possible for an especially large female to outnumber egg production of all the other females combined. If this is the case, the biggest female has a greater chance of producing more offspring by staying female and letting the next-in-line female transition into male. This is especially true when there happen to be a lot of other males hanging around the periphery of the harem, streaking in and adding their own genes to the mix. These invading sperm donors increase sperm competition for the dominant male, likely reducing his fertilization rates. On the other hand, the extra sperm may boost fertilization success for the female.

Because even a small male has enough sperm to fertilize all her eggs (and those streakers may up the sperm count, too), and she has more eggs than all the other females combined, the BOFFFF reaps higher reproductive reward by staying female. How bucktooth parrotfish became such discerning accountants remains a mystery, but it likely has to do with visual cues about the size of harem mates and presence of other males swooping in on the spawning.

In other species, it's not so much the numbers but the nearness of the neighbors that sparks the urge to transform into the other sex.

A DIFFERENT KIND OF PEER PRESSURE

The Johnsons liked their quiet street. Their beautiful house. Their lovely backyard. Life was good. That is, until the Smiths arrived. That's when Mr. Johnson started hanging around his garden plot a whole lot more, admiring groups of gals Mrs. Smith was constantly entertaining. At first, Mrs. Johnson didn't raise a fuss. But as Mr. Johnson continued to linger by the driveway longer each day, she grew uneasy. Never before had there been so much temptation so close to home.

Mrs. Johnson couldn't help but feel neglected. With Mr. Smith away at work, Mr. Johnson began to unabashedly enjoy the company of his new lady friends on a daily basis. After a few weeks of this behavior, Mrs. Johnson began making some changes of her own. She avoided Mr. Johnson during the day. When they did cross paths, she wanted nothing to do with him. She was combative and easily agitated when he did try to talk with her. Mr. Johnson grew suspicious. What was up with this aggression? Where was she going at night? He started to wonder if there was another man in the picture. But there wasn't. At least, not yet.

Then, about a month after the Smiths' arrival, Mr. Johnson awoke to find his wife sprouting a stubbled chin and wearing his best pair of suit pants, packing up the house. "I'm leaving," she said in a deep and rather shocking baritone. "And I'm taking the other girls with me." And with that, she hoisted the flat-screen TV off the wall and onto her shoulders (when did they get so broad?), loaded the car, and drove off, Mrs. Smith and her girlfriends waving out the back window as they disappeared around the rose bushes.

Mr. Smith, standing in his robe and looking rather disconsolate, slowly walked over to where Mr. Johnson stood, despondent, in the driveway.

"I thought there was another man," said Mr. Johnson.

"There was," said Mr. Smith. "It was your wife."

"I hate it when that happens," said Mr. Johnson.

————

THE FEMALE RUSTY ANGELFISH is anything but rusty at seducing and absconding with several members of her own harem when the reef starts to get crowded. During the breeding season, a male tightly guards one to six females, courting all his ladies for about thirty minutes before sunset. Then, one by one, each will join him in an upward spawning rush as the evening commences. It's quite romantic, as long as there aren't any other rusty angelfish harems around. Should one or more harems pop up in the neighborhood, though, things get a little complicated.

Males will strategically visit nearby harems when the dominant male isn't looking (or perhaps is off visiting another harem himself). The more time he spends with neighboring females, the less time he has to socialize and court his own harem. Especially for the largest female, this lack of adequate attention can result in the cold shoulder . . . and a potential coup d'harem. She starts by simply shunning his evening advances (hello, Mr. Johnson!). Such denial strikes deep: in the dark, celibate night, abstinence makes the fish grow longer. By forgoing sex, a rusty angelfish gal can shift energy from egg production to growth and gain length on the resident male. A few weeks later, she bursts forth in all her masculine glory and swims off with a number of her former harem mates. The more harems concentrated in a local area, the higher the likelihood that such antics will take place.

Whether it is via careful calculation or lack of adequate attention, the idea that the number of males or females in a region could impose a change to your own sex is completely foreign to us. But sex-change-by-peer-pressure is a major driver of successful sex beneath the sea. It also happens to be a reproductive strategy particularly at risk to overfishing, especially when the biggest fish are targeted.

The fishing industry tends to be size-selective, with gear (or divers) that often target certain individuals in the population. This is a good thing when it helps fishers avoid catching juveniles who haven't yet had a chance to reproduce. It can also lead to better income, as the biggest fish or oysters in a population often fetch the highest prices at the dock. But going after the biggest fish in the sea when those are all the males, for instance, can significantly skew the sex ratio of a population. This is bad news for future generations as it makes it harder for females to find males, and even when they do, if sex ratios are really skewed, there may not be enough sperm to go around.

In protogynous sex changers, the longer a female waits until she changes sex, the bigger she grows and the more eggs she can pump out. If fishing selectively removes the biggest males, females change sex sooner in order to compensate for the loss of males, but at a cost: changing to males at smaller sizes than they otherwise might deprives the population of big females. In other words, there are fewer BOFFFFs to help restore the numbers lost to fishing.

On the other hand, in those species that switch from male to female, targeting the biggest individuals is kind of like killing the goose that lays the golden egg. It's not a smart move if you want to keep seeing your favorite shrimp or oyster on the menu.

This kind of unnatural size selection affects more than just sex-changing species; it can even drive genetic changes in a population and cause . . . shrinkage. This can happen irrespective of the sex-change capacity of the species. To date, selective fishing pressure has led to reductions in the size, age of maturation, and growth rates in wild populations of fish and oysters—characteristics that are important for making strong, healthy adults. Fast growth and larger size are great for competing in nature, but they are the kiss of death under selective fishing, so these traits end up being weeded out from the gene pool. This leaves a population of diminutive survivors with features that don't bode well for future generations. In experimental studies on Atlantic silversides, an important prey species along the Eastern Seaboard of the United States, researchers found that selective fishing for the biggest fish resulted in smaller males and females that produced fewer sperm and eggs and had slower-growing larvae. These individuals are less likely to become the "big fish" that we would scoop up, but they are more likely to be gobbled up by natural predators.

Compared to their ancestors, the offspring of fish from such size-selective fisheries can be less fit for surviving the challenges Mother Nature imposes, which makes population recovery a painfully slow process. In the Atlantic silversides, the experiments showed it can take more than twice as long for a population to recover its prefished growth rates and sizes as it took to lose them. Similar effects may be happening with eastern oysters in Virginia. Fishing pressure and disease have reduced life expectancy from ten to twenty years down to about five years. This pressure may be the reason why some oysters appear to mature earlier and undergo sex change at smaller sizes than previously documented. The good news is the studies on the silversides show that recovery is possible, as is shrinkage reversal. It just may take a while. Though perhaps it goes without saying, it is thus best to avoid shrinkage in the first place, whenever possible.

FLIP-FLOPPERS: SPECIES THAT
CHANGE SEX MORE THAN ONCE

There are some even more extreme responses to peer pressure in the deep—responses that are far less deterministic than those explored thus far. For some species, sex change is a two-way street.

The polychaete (rhymes with jolly-feet) worm, *Ophryotrocha puerilis,* is less than an inch long, often transparent, and excels in a form of sex-swapping leapfrog. Like clownfish, these worms form spawning pairs, with the larger worm performing as the female, so as to take advantage of the "larger size equals more eggs" phenomenon. But because males grow more quickly than females, the male soon outsizes his mate. Unlike clownfish, the female doesn't bully the male into quiet subordination. Instead, she welcomes his growth spurt. Upon reaching the bigger size, the male then switches to female as she undergoes a sex-change *reversal* back to male. She-now-he will then gain the growth advantage, catch up, and eventually surpass the length of her partner, and the two will swap roles once again. At any time the pair benefits from having the largest individual play the role of female while taking advantage of the enhanced growth rate of the male role.

Such advanced flip-flopping isn't limited to crafty worms, however. Several species of fish are known to have the ability to reverse sex change as well, but perhaps none quite as deftly as the bluebanded goby. This small, bottom-dwelling fish lives in the shallows along the west coast of North America and is a favorite of kelp forest divers. About the length of your little finger, it looks like a cross between a Swedish Fish candy and a Mexican wrestler, face mask and all.

With neon-blue vertical stripes banding a bright red-orange body and blue markings across the face, this is *the fish* when it comes to sex change, according to Dr. Matthew Grober, associate professor of biology at Georgia State University: "Males change to females, females change back into males. Give me a week and I can teach a seventeen-year-old freshman how to get these fish to change sex. They [the gobies, not the freshmen] are that good at it."

Though they appear rather nonchalant in their sex-swapping escapades, such elasticity of sexual expression is actually governed by strict social convention; it's about as carefree as a convent. Which is why Grober says anyone can successfully manipulate the conditions and cause sex change in either direction, once you know the rules.

As transparent little larvae, the sex of these gobies is undetermined: their gonads and genitalia are a mix of male and female. After hatching, they swim near the surface for a few months in this ambiguous state, and then as juveniles settle down to live among the rocks and other gobies on the seafloor. Depending on where they land, social rules will dictate whether they develop into females or males. If there is a male around exerting dominance, the juveniles become female; if not, then one of the new arrivals will start acting more aggressive, taking on the male role. It's not clear what causes one fish out of a group of newly settled youths to be so bold, but once one starts developing as a male, he gains a growth advantage over the other juveniles, which reinforces his superior status. Other juveniles or females in the area will remain subordinate. They stay females as long as another fish is there to exert dominance. In this manner, bluebanded gobies set up small harems of three to ten females per male.

As life goes on in a harem, the male has his work cut out for him. First, he has to tend and defend the nest, even from his own females— left unchecked, females will eat their own eggs. Second, he's got to court and mate with the females in his harem. Third, he must defend his territory against other males. With such a full plate, he designates a second-in-command: the alpha female, the next most aggressive in the group, to dominate the other females. This kind of bullying can keep other fish in the harem from growing too big and threatening the leading pair's social rank. All a male bluebanded goby has to do to control his harem, then, is dominate one female and let her take care of the rest. There are risks to this strategy, though. A too-aggressive female can stunt egg output in her fellow females, and that's not good for the male. It's a rather complicated balancing act, but such is the price of unfettered access to multiple females. Should the male die, or for whatever reason fail to exert such dominance, the highest-ranking female will quickly transform into a male and take over the harem. If this sounds familiar, it is because it's the same tactic clownfish use, only with the opposite sex at the top. But here is where blue-banded gobies get a bit crazy.

A high-ranking female in the absence of a male can change sex and take over the harem. But if another, more dominant male later enters the picture, the female-turned-male has three options: try to defend the territory; bail out and go find a harem he can take over; or opt to shift from male back to female. The advantage here is that, if the odds don't look good for winning the battle or overtaking a different harem, the smaller male can continue to reproduce as another female in the group rather than lose the harem and all reproductive potential.

This sexual flexibility means that when two similar-ranking adult fish of the same sex meet, either one of them could switch to form a complete pair—it takes about three weeks. As with juveniles, a not-yet-understood cue triggers one of the two fish to assert dominance. In doing so, this individual will gain the growth advantage. If it is two males that meet, the aggressor will keep his male status; the other subordinate fish switches sex. If it is two females that meet, the aggressive one will morph to male. This happens every time.

And when life gets really complicated, these little risqué rompers do something rather extraordinary: they *un*-sex.

Reversing maturity, they can revert back to their juvenile ambiguous state, precursors to either gonad at the ready. Grober discovered this trick of the trade when he entered a group of bluebanded gobies in a horrid speed-dating marathon in his lab, introducing one fish to a new group every day. In so doing, Grober confused the signals of social status. And when rank is unclear, so is the goby's sex life, and thus their body parts. As Grober puts it, "They go back to vanilla." In other words, sex in these gobies is determined either through subjugation by an aggressor or by rapidly asserting dominance over less aggressive reef mates. In the absence of any indication of which role to play—dominatrix or submissive lover—an individual waits in bi-sexed limbo. Talk about shades of gray.

Being so sexually labile has its risks: a world where gobies flip-flop their sex left and right, day and night, all willy-nilly, would surely sink the species—nobody would know who was supposed to do what. So although they have extreme flexibility in their sex, the gobies don't bend the rules when it comes to swapping roles. They follow a strict, clear social convention: if subordinate, be female; when in doubt, wait it out.

Not all sex change in the animal kingdom is part of the natural life cycle of a species. For many animals, the presence or absence of other species can alter one's sexual destiny. Imagine if the squirrels in your backyard had that kind of influence on you—and now, read on.

INGESTING AND INFESTING:
THE INFLUENCE OF PREY, PARASITES,
AND POLLUTANTS ON SEX CHANGE

Diatoms are microscopic, single-celled algae that float through the open sea converting sunlight into energy and fueling the ocean food web. They also fuel us. Collectively, diatoms pump out about 20 percent of the global oxygen supply. Though you may not have heard of them, you can thank them for at least one of the breaths you took while reading this page. Master alchemists, diatoms capture silica floating in the water to build glistening shells called frustules. Like a vitamin capsule made of glass, the two frustule halves fit snugly together, creating a protected chamber for the soft parts of the cell. Under a microscope, diatoms transform a droplet of saltwater into a surreal galaxy: centric diatoms float by like translucent lemon slices; others, called pennate diatoms, are elongate, resembling crystal rocket ships that spin and thrust across a slide.

Where diatom wizardry truly excels, however, is in chemical warfare, a talent you may not think you know about, but probably have seen or heard tales of—especially if you are a Hitchcock fan.

Domoic acid—a neurotoxin produced by some diatoms—accumulates in filter-feeding fish and shellfish that suck up these plankton as food. People who eat infected bivalves, such as clams or oysters, may experience short-term memory loss and digestive problems and, in rare cases, they can die. Domoic acid poisoning can also affect other animals that ingest infected prey. Such was the case one summer morning in 1961 when thousands of sooty shearwaters, a small and normally docile seabird, bombarded the seaside town of Capitola, California, crashing into windows and lamp posts, and attacking people in the streets. The freak ambush caught the attention of Alfred Hitchcock, who often vacationed nearby, and helped serve as inspiration for—you guessed it—*The Birds*. It is now widely believed that the shearwaters, known to dine on small filter-feeding fishes such as anchovies, went batty due to domoic acid poisoning.

Not all diatoms seek to unleash death and insanity on their fellow ocean dwellers, though. Some aim to simply transform them. For one species of shrimp, eating too many diatoms results in early onset of sex change to female form. Normally, when no diatoms are around, the shrimp undergoes the typical transition of maturing first as a male, and about a year later transitions to female. But when larval shrimp gorge on these diatoms in the first few days of life, their food unleashes a chemical cocktail that stops male development cold and instead triggers development of female gonads and genitalia. The compound responsible for this remarkably focused attack remains unknown, as does the reason for it; it is unclear how diatoms benefit from more female shrimp—perhaps they don't. Maybe the effect is just a by-product of the sexual flexibility in the shrimp: the chemical compound could have evolved to do something else in another species, or even serve a yet unknown role for the diatom itself. Either way, such sex-change-by-consumption shows how malleable some species really are when it comes to their sexual status. And it's a flexibility some particularly crafty species have learned to exploit.

Dr. Alex Ford, a researcher at the University of Portsmouth in England, gets especially enthused when trying to work out how sex determination "goes wrong" in a species. And for the past few years, something was definitely going wrong along the Scottish coast with a local species of amphipod, a small crustacean that looks a bit like a shrimp that got stuck between two elevator doors—flattened on both sides. A critical food source for water birds and fish, this one species of amphipod normally lives in balanced 50:50 sex ratios and is not a sex changer.

Ford has found that in some places the sex ratio is skewed—highly skewed—with up to four times as many females as males. He's also discovered many male amphipods with female characteristics and an unusual number of females with male parts. If you are wondering how to tell a female from a male amphipod, it's not too hard. The males have a small pimple-like protrusion, called a papillae, while the females have a modified appendage on their leg that they use to hold eggs. The "intersex" individuals Ford has found often have both. In addition, their inner sex parts—the gonads—may also be scrambled.

The high numbers of females compared to males provided Ford a clue as to what could be going on. In the crustacean world, parasites are known for far more than nasty stomachaches—infestations can cause sex change. It's all part of the parasite's ploy to spread its offspring using a very effective tactic called vertical transmission.

We tend to think of parasites transferring from host to victim via blood or skin contact, but not all parasites do that. Some transfer their larvae via the developing eggs of a female host. It's pretty genius (from the parasite's perspective) if you think about it: the parasites infect the next generation of hosts (the eggs) before they ever leave their mothers' bodies. Their victims are born infected. But there is a potential major glitch: a parasite that infects a male amphipod has reached a dead end—no available eggs to transport the parasite's offspring. To avoid such abrupt termination of the line, parasites that use vertical transmission often employ the nifty trick of turning males into females.

When Ford looked closely at amphipods from different sites around the coast, he found a parasite called microsporidian that is known to be able to alter the sex of their hosts, and not just by converting males to females. Once inside a fertilized egg, the replicating parasite can force the egg to develop as a female, skewing the sex ratio of each successive generation of hosts toward more and more females.

Ford actually found two kinds of parasites associated with the amphipods, but the second kind isn't known to sex change its host. What effect it does have, Ford is still trying to figure out. Yet parasites are only *part* of the sex-change dynamic. Ford explains that sex in amphipods, like many other crustaceans as well as reptiles and fish, also can be affected by the local environment. Water or sand temperature—as noted in the case of sea turtles—the length of sunlight during the day, or amount of food in the water, can equally alter final sex determination in many species of fish and crustaceans. So can pheromones, as is the case with the towering stacks of slipper shells and even oysters. Sometimes these chemical cues are natural, but increasingly they are not. Pollution in coastal and open seas is on the rise and increasingly impacts marine life, especially in species with more impressionable sexual systems.

Previous studies have shown that common contaminants in the environment, especially what are known as endocrine-disrupting chemicals (EDCs), can cause feminization—the adoption of female characteristics by genetically male individuals—in a range of marine species, from crabs to mollusks to sea urchins. Perhaps the most infamous example is that of the worldwide epidemic of penis-growing female snails back in the 1990s. In that incident, female snails began showing a disturbing condition: their oviducts were blocked by anomalously sprouted schlongs, causing reproductive failure and death. Not surprisingly, many snail populations crashed, taking down with them entire marine communities that depended on the snails for food. After much debate regarding the cause, eventually studies linked the deformities to high amounts of tributylin (TBT), a compound in boat-bottom paint and a virulent EDC.

EDCs work by "impersonating" natural hormones and disrupting internal hormonal cycles. They include bisphenol A (BPA), phthalates, and polychlorinated biphenyls (PCBs), among many others. These compounds are found in everything from industrial by-products to consumer goods, pesticides, personal care products, and birth control pills. They literally flush into waterways via discharge from sewage treatment plants, which don't filter out the molecules.

Across the different regions Ford has studied, he's found the highest numbers of intersex amphipods in sites regularly bathed by industrial effluence from nearby petrochemical factories, shipyards, and pulp mills. Although he has yet to identify a specific contaminant, there seems to be a link between pollution and genital mix-ups. However, the high-pollution sites also have the highest rates of parasite infection. Ford is still trying to tease apart what insults (parasites vs. pollutants) are affecting which amphipods in what ways. It is possible that the parasites might suppress amphipod immune systems, making them more susceptible to effects from certain pollutants; or it might be the other way around, with amphipods in contaminated environments unable to fight off parasitic infections that then cause sex change. After fourteen years of studying this stuff, Ford has more questions than answers. One thing all this complexity does prove, however, is that the consequence of nature plus human activity is not easy to predict.

SEALING THE DEAL

Part 1: Sexual Intercourse

THE ENDLESS, TREACHEROUS SEARCHING, THE IN-
tense competition and fierce battles, the artful seductions and sex changes,
the effortful foreplay—it is such a long journey and still the deed is not yet
done. Even once a potential suitor navigates all of that build-up, the final
test comes down to the last few inches (or feet) between the meeting of the
egg and the sperm.

These next few chapters explore the myriad ways sperm make it down
the home stretch, actually transferring from the male to the female—or
her egg. For some, this process is an intimate affair, for others, a more
fulfillment-from-afar approach, with sperm and eggs shooting off into the
swirling sea to sort it out for themselves. This latter approach is unique to
a liquid environment—and likely the way sex first began. (Think about
that the next time someone winks at you from across the room.) But we'll
start with a more familiar method of achieving sperm-egg unions: deliver-
ing the sperm into or onto the female body, courtesy of a funky appendage
known in the science world as an "intromittent organ." You know it as the
penis, in human history responsible for all manner of anxiety disorders,
decades of misguided psychoanalysis, and continued bad car design. But
its role in marine life reproduction is even more interesting.

— 4 —

THE PENIS CHAPTER

Sex as a Contact Sport

SEX-SEA TRIVIA

- *Snails can ditch and then regrow their penis.*
- *The longest penis (proportionally) is eight times the length of the male's body. Guess who owns it?*
- *There is a species named "astounding swimmer with a large penis."*
- *The male argonaut has a detachable, projectile penis.*
- *Penis-fencing is serious business—just ask a knocked-up flatworm.*

SEX-SEA SOUNDTRACK

1. "The Penis Song"—Monty Python
2. "I've Got You Under My Skin"—Cole Porter
3. "My Ding-a-Ling"—Chuck Berry
4. "Sledgehammer"—Peter Gabriel

JOHNSON. DONG. SCHLONG. MEMBER. UNIT. TACKLE. Cupid's fiery shaft—a penis by any other name would function just the same. Such extensive and colorful penis lingo suits the enormous diversity of copulatory structures wielded by males from barnacles to whales. A quick scan of oceanic penes (rhymes with weenies and is the proper and rarely used plural of penis) takes us from the long and thin to the short and fat, from immobile to prehensile, from hard to soft. There is no such thing as a typical or even "average" penis. This chapter is all about celebrating that uplifting fact.

Male genitals, like their female counterparts, are some of the most diversified structures on the planet. While generally designed for depositing sperm, the penis participates in activities far beyond simple drop-off duties: they may physically scrape out sperm from previous males, or entice the female to accept insemination through mechanical or chemical stimulation; they may damage the female in a way that prevents future matings; and they may serve to sneak in a copulation here or there when a female is locked in the embrace of another lover.

The main purpose of a penis, however—besides being an exit hose for urine in us mammals—is to allow for internal fertilization, or in some cases such as squid, *nearly* internal fertilization. From the male's perspective, a phallus enables direct deposit of sperm as close as possible to the egg and within a relatively confined space. This helps boost sperm concentration and offers benefits to the developing embryos, including protection from predation and a controlled environment optimized for growth.

Females may also benefit in that—before and after sex—they can screen for suitable suitors and sperm. This isn't possible when spawning happens externally, where wave and tide create random encounters between sperm and egg (though eggs may do some selection on their own, but we'll get to that later).

For both males and females, sex-by-penis offers greater control. But it also complicates copulation (and life itself) in all sorts of ways.

The fact that any sort of penetrating (or even semipenetrating) sex organ succeeds as a sexual strategy is quite surprising: many penises burrow pointy, horned heads way up inside a female, introducing risk of infection and causing physical damage. Foreign, invading bodies normally pose quite the threat to the recipient (think parasite). In addition, though some sea life takes a "hit it and quit it" approach, sex-by-insertion in many species requires extended holds in intimate, if not awkward, positions, making the pair of lovers more vulnerable to predators during their interlude.

Just ask the male queen conch.

In the sunlit shallows of the Caribbean, this iconic snail may live thirty years and, in that time, grows an impressive foot-long spiraled shell. Living a normally independent life, males and females heave-ho these hefty mobile homes along the seafloor, munching algae and sea grass along the way. When the time is right, they gather to mate, sometimes in large groups. But with gigantic, armored, calcified bomb shelters on their backs, close contact is rather challenging. At least it would be, if not for the male's impressively expandable penis, known as a verge.

An active verge looks like a startled earthworm: about the thickness of a pencil and dark black, it wiggles its way out from underneath the male's shell and across the sand to slip in underneath the small, domed archway at the rear of the female's shell. This is the doorway to her inner chamber and genital groove, where fertilization takes place.

But it's a dangerous trek: stretching nearly half the male's own body length, the verge is easy pickings for patrolling crabs and other predators. Happen upon a mature male conch and you're likely to find he's nursing a partial penis, clipped off in the line of sexual duty. The damage is not permanent, though. In time, conchs, like many other mollusks, can simply regrow their penises.

And while conchs must contend with crabs to preserve their penis, other species use their members as weaponry in elaborate duels to the reproductive finish.

FENCING PHALLUSES

Like the dry, vacant roads of the Old West, long stretches of sandy floor spread as endless plains fading into a pale horizon. Stirred by unseen currents, sand plumes rise like dust devils, swirling. A tangle of seaweed tumbles by. You can almost hear the far-off, forlorn whistle, harbinger of the showdown. Slowly, you begin to climb the side of a sand ripple, as steep as a mountain before you. And there, at the crest, your senses begin to tingle.

On the horizon, from the faded blue distance, your opponent emerges. Descending the hill of sand, you confirm that this indeed is a worthy adversary. The distance between you narrows and you prepare for the assault. Rising up vertically, like the face of a wave, your head crests forward and you reveal your weapon. Forged by nature alone, this battle will not be fought with steel or iron, but with a male's most prized possession, his most precious appendage: his penis.

SPAGHETTI WESTERNS would look mighty different if cast with flatworms rather than cowboys.

In the marvelous marine realm, there is no substitute for what Mother Nature has bequeathed. Two challengers simply rear up and clash their mighty man-swords to assert dominance and, with any luck, successfully stab their opponent.

Unlike human duels to the death, however, the objective of this battle is not to kill. To the contrary, it is a fight to spark life.

Nature's most dexterous of fencing matches takes place between two animals each as flat as a pancake, each one bent on inseminating the other. They are penis-fencing hermaphroditic flatworms, found in shallow tropical waters such as the reefs off Australia. For most of their lives they move slowly, gliding gracefully over the seafloor, undulating ripples along their outer edge propelling them forward. But when opportunity strikes, so does the flatworm.

In an impressive display of athleticism, these mostly two-dimensional creatures rise up like king cobras. Balanced on their rear halves, they peel up off the sand floor, exposing a clear white double-headed penis appropriately termed a stylet. They sway back and forth, two wrestlers before the grab, looking for the opening. Without bones, they flex and bend at extraordinary angles, as fluid as their environment, limited only by their marginal length. Many strikes will miss, or be aborted at the last minute as one flatworm ducks and dives to avoid being stabbed. Their thin bodies ruffle, flexing with exertion, coiling and striking, curling and bending, flattening and gliding into a new position. These worms match agility with endurance, engaging in battles that can last up to an hour, with both combatants likely suffering stab wounds by the end. It is a rather violent mating strategy for species that technically have all the parts they need to perform a much gentler self-fertilization, so why go through such an ordeal?

For starters, sex with oneself isn't always a mild affair. In flatworms eggs and sperm may be stored in separate compartments inside the body, which means self-fertilization requires a bit of gymnastics in order to get sperm to meet egg. A recent study on a transparent species of flatworm called *Macrostomum hystrix* showed that such maneuvers can include stabbing oneself in the head. Kept in isolation, these loner worms perform extremely contortionist self-fornication. With the stylet located near the tail, they somehow flip their bottom half up to meet their top, jabbing themselves right in the noggin. From there, the sperm presumably just wiggle their way back south toward the ovaries where they inseminate the eggs.

Lucky for the worms, this stunt is a desperate measure reserved for desperate times only. When presented with the opportunity to mate, the see-through worms stab each other in the tail region—no self-inflicted head piercings going on. Turns out, the impressive folding trick is not the preferred way to sow one's seeds. In general, if given the chance, most hermaphroditic worms, the aforementioned extremely bendy bedfellow included, choose not to self-seed. Self-fertilization appears to just be a way to put a puncturing penis to good use in hard times—but it's not likely the reason such pointed phalluses evolved; we'll get to that theory in a minute.

Overall, mixing genes, as good old-fashioned sex is designed to do, remains the preferred method for reproduction, probably because it brings the benefits of diversity while avoiding the side effects of inbreeding. A hermaphroditic existence simply means you don't have to be so picky with choosing your partner: any sex will do. A pair of hermaphrodites need only decide who plays which role.

Some species rely upon a reciprocal arrangement, where sperm transfer occurs simultaneously between both individuals. In sea hares—bulbous snails without shells—species can form enormous sex chains by interlocking their forward-located penises into the rear-located vaginas of the sea hares in front. In some Mediterranean flatworms, pairs of copulating worms simultaneously insert male parts into the female opening part, forming a sort of sex loop. In such reciprocal sex, the male organ enters the genital opening in the same manner as sex between male and females, but both partners give and receive all at once.

But in species such as penis-fencing flatworms, individuals skip the vagina and land their cargo anywhere on the body. Scientists hypothesize that the evolution of insemination-by-stabbing is a way for the sperm donor to bypass the female's reproductive tract and avoid any kind of defenses within it, such as sex plugs left by previous mates, or features of the female's genitalia that may selectively remove or favor some sperm over others (more on that in the next chapter). Instead, these species use their stylets to deposit the goods anywhere on the body and sit back as the sperm swim their way to the site of fertilization.

A flatworm from off the coast of Australia takes things one prick further. In 2013, researchers watched as pairs of *Siphopteron sp.* repeatedly brandished their enormous, forked phalluses and used them to both inseminate *and stab* each other in the head. One tip of the split penis carried the sperm; the other, with the more pointed stylet, transferred a secretion produced by the prostate. This was delivered directly into the forehead of its mate. Why the focus on the frontal lobe? We still have no idea. With penises equal in length to their bodies, the worms could aim anywhere; yet every copulation included a jab right between the eyes.

From reciprocal sex to traumatic puncture techniques, these approaches all share this in common: they minimize the costlier, female side of the sex equation. In almost every type of mating situation, females wind up carrying the burden of growing energy-expensive eggs and caring for the young. The males, on the other hand, get off with little investment (those dutiful egg-guarding dads are an exception). For hermaphrodites, any mating encounter is an opportunity to inseminate or be inseminated. It's far better, energetically, to be the inseminator. But that position isn't always easy to achieve.

In truly reciprocal species of hermaphrodites, such as some sea bass, individuals agree to play both roles but need to ensure nobody cheats. So, they employ an "I'll scratch your back, you scratch mine" strategy: one partner releases some eggs while the other puffs out some sperm, then they switch roles. They may repeat this pattern, each individual alternating between spawning some eggs followed by some sperm as their partner spawns the complementary gametes. In these arrangements, each partner is willing to exchange some female burden for the opportunity to also fertilize some eggs.

Other approaches are far more devious. In the Australian head-stabbing flatworm species, researchers speculate that the released secretion may force the impaled mate to more readily retain the sperm. Such crafty chemical cocktails are common in ejaculates, including in houseflies. The jury is still out regarding what exactly the flatworm cranial injections confer, but it's likely got something to do with boosting success of sperm.

Such self-serving strategies are common. Even flatworms that appear to exhibit more "cooperative" mating systems aren't always as civilized as they look. Take sex-loop-forming Mediterranean flatworms, for instance. Each partner invites the other to penetrate. But after the couple parts, these flexible flatworms fold in half, put their mouths on their genitals, and attempt to suck out the freshly deposited sperm. At least they do seem to wait until the other isn't looking.

Such vacuuming up of seed isn't easy, however, given the spiky heads of the sperm—an adaptation that likely evolved as a response to counter such oral assaults. In species where males puncture females all over using a pointed penis, their sperm are far smoother and simpler in shape, making it easier to glide through the body to the eggs.

Overall, the penis-fencers take the most selfish approach, with each individual insisting on performing as the male and battling to land the perfect shot. In some species, stabbings are not equal—sperm must be deposited in the right place to make it to the eggs. Thus, a well-placed poke can allow one flatworm to inseminate while avoiding the other's stab and escaping the burden of caring for fertilized eggs.

Few species wield their penises with such agility. In fact, the first phalluses to evolve were far more rigid.

ANCIENT APPENDAGES

Ancient penises and their typical modern marine relics tend to be hard. Really hard. The males of a prehistoric fish called antiarchs sported a pair of hooks made of solid bone that extended out of armored plates. These "claspers" slotted sideways into very rough-hewn genital plates (one paleontologist refers to them as "like cheese graters") on the female, which acted like Velcro to lock the male's unit into place. Using tiny arms to do-si-do themselves together, the two fish copulated in a square-dance-style sex position. Discovered in late 2013, this finding pushes the known origin of internal fertilization in vertebrates back to 385 million years ago. Which is old. But the dawn of the first known penis of any animal precedes this by about 40 million years, to 425 million years old. That's the age of the oldest known fossil penis—a johnson that left quite an impression.

Like most tales about penises, this one involves a massive explosion. An ancient volcano near what is now the United Kingdom sent ash shooting skyward. It drifted down slightly east of the Welsh border, settling into what was then a coastal sea. Oblivious to the disturbance above the surface, thousands of marine dwellers were suddenly smothered in the fine dust, which quickly hardened around their bodies. They were trapped, encased within skin-tight molds that hugged each curve and filled every crevice. Thousands of fossils formed, time capsules of animal life from a long-lost seascape.

These capsules remained sealed, impossible for paleontologists to decipher; to break them open was to destroy their delicate anatomy. Using a new technique, however, scientists today can grind extremely thin slices of these rocks—twenty microns at a time (about one-fifth the thickness of a strand of hair)—and snap a photo of the imprint beneath. Researchers then use specialized software to stitch these images together into a three-dimensional picture of the animals that lie within. These extremely detailed reconstructions reveal antennae and feeding appendages, segmented body parts, and compound eyes—and, in the case of one truly remarkable specimen, the existence of a very conspicuous phallus.

Just how conspicuous? Enough to inspire paleontologist Dr. David Siveter at the University of Leicester and his colleagues at Oxford, London, and Yale, to name it *Colymbosathon ecplecticos*—from the Greek for "astounding swimmer with a large penis."

Over a decade after his extraordinary find, Siveter is still in disbelief that he can lay claim to being the discoverer of the world's oldest penis. Even for a weathered paleontologist, finding a fossil penis—or any soft parts of an animal—is extremely rare. They just don't preserve well. But these fossils were preserved so well the researchers could make out the delicate limbs, the eye, the gills, and the copulatory structure—and it was an enormous penis.

Siveter had been studying ancient ostracods for decades when he made his discovery in 2003. Today, one of the eight thousand species of ostracods are as likely to be swimming around the pond in your backyard as they are scooting through the deep sea. Picture a shrimplike critter squashed between two lima beans. But the lima beans are thin, often transparent shells, and a full-grown adult is likely to be from the size of a poppy seed to about as big as a ping-pong ball. Two fine antennae, one used for swimming and one for sensing the environment, stick out like tiny hairs from a small opening in the shell. Numerous and successful crustaceans in modern times, they are also the most ubiquitous arthropod in the fossil record—their tiny hinged shells are everywhere. But until Siveter's find, paleontologists had never actually seen traces of the animal's body in ancient fossils—and certainly never a penis. Here in this one fossil, they had undeniable proof of the oldest known male. The arthropod Adam.

Thanks to this discovery, we now know that living ostracods come from a long line of well-endowed crustaceans, a trait that's been remarkably conserved through the ages. When discussing the modern species of today's oceans, lakes, streams, and ponds, Siveter notes that some males have copulatory structures over one-third the length of their bodies.

That's like having a six-foot-tall man sport a two-foot-long penis. Only he would have to cart around two penises that length: an ostracod phallus comes as a pair.

Siveter explains that the copulatory structures, called hemipenes, possibly evolved from what used to be a set of limbs, so there is one on each side. This makes ostracod males righties *and* lefties. And they use these dual appendages to pump out sperm of gigantic size relative to their own bodies: in at least one species of living ostracod, the sperm are *over ten times the length of the entire male.* So that same six-foot tall guy with the two two-foot-long penises would let loose a single sperm longer than a school bus.

Here's a tiny one- to two-millimeter-long animal and it's making sperm much bigger than an elephant's. And scientists still don't really know why. To manage such enormous wrigglers, a male ostracod has to knot and wind the trailing tendrils into tight coils and then shoot them into the equally tiny female, one through each penis. He's evolved a specialized structure that works like a bicycle pump to achieve this task; she, in return, has two vaginas to receive the associated spaghetti balls of sperm.

When I ask if the males alternate, using one penis at a time, Siveter says no: "It's not shift work."

In other words, ostracod sex is like firing a double-barrel shotgun, both penises simultaneously releasing humongous sperm into the two genital openings of the female. Remarkably, this double-penetration doesn't require a specific docking approach. A male can mount a female in many positions, including belly to belly, or from behind. The result is a kind of creative crustacean Kama Sutra. She just needs to open her shell to let him in.

But that's a big "just." Ostracods can clamp their shells so tightly they can survive the digestive acids and pass right through the guts of predatory fish. If she is not up for it, there is no way he's getting in. Male ostracods appear to need consent. Which may explain why some males go through remarkable displays in order to woo their mates.

Every night, just past dusk, the upper water column of the Caribbean Sea alights with iridescent blue bursts that flicker, streak, and fade against the black waters. The source of this marine meteor shower is thousands of male ostracods squirting liquid light out of glands in their heads. The chemical reaction works like epoxy; two compounds produced by the ostracod react when mixed with seawater, emitting a bright blue light that is used by the males to attract females. Ancient and accomplished, each species has its own pattern of sparks, a fleeting Morse code of dots and dashes. Leaving their bottom abodes to rise to specific depths at specific times, the males proceed to paint the water column with pointillist pyrotechnics that turn the blank, black waters into Van Gogh's *Starry Night*.

If effective, the displays capture the attention of nearby females who follow the shimmering trails to find a worthy mate. But the blackness between the bright bursts masks lurking interlopers, males waiting to intercept unsuspecting females. Whether she reaches the target male or finds herself ensnared by a sneaky suitor, the result will be the same: penetration by an impressive pair of penises.

In most species of ostracod, the female deposits the fertilized eggs on the seafloor or on some other surface. In some species, females brood the eggs, and in a select few, hatched juveniles continue to develop inside the internal brood pouch of the female until they can better fend for themselves. Although such extended parental care is unusual, it can now be dated back 450 million years, with the finding of a new species of ostracod. Siveter and colleagues named this one *Luprisca incuba,* which means "ancient egg-brooding mother." The fossil is from upper New York State and shows an ostracod female preserved with eggs and hatched larvae still inside her. She is the arthropod Eve to complement the ancient Adam.

PLIABLE POKERS

At some point, penises left the permanently rigid path and diversified into a plethora of forms. The evolution of increasingly flexible members required enormous adaptations. It takes some serious engineering to evolve something that is hard enough to penetrate but durable enough to endure repeated thrusting. More bendy boners allow their hosts to copulate in more positions and hide the family jewels when not in use—something that comes in especially handy when trying to maintain a hydrodynamic (streamlined) form in the water.

Many fish do not have this option. Their sex structures stick out at sharp angles from the body, which can be a real hindrance when trying to escape predators. Guppies and their cousins the mosquitofish, for example, are one of the few fish with internal fertilization, made possible by an extended anal fin ray called a gonopodia. Female mosquitofish find bigger gonopodia more attractive, but these enlarged packages slow a fish down, making the more well-endowed males easy pickings for hungry predators.

Marine mammals, on the other hand, have adapted penises that can be tucked away, aiding with swimming speed and temperature regulation—a necessity when you spend a lot of time in frigid waters or lying on the ice. Nobody wants a popsicle for a penis. Hideaway penises were made possible by a softer approach to penetration, something mammals had already achieved on land in one of three ways.

Humans, along with horses, armadillos, and a few nonmammalian species including turtles, elegantly achieve effective erections with an inflatable penis—a technical term, by the way. These male members rely on hydraulics to pump spongy tissues full of blood in order to stiffen. And here's a fun fact: in a cleverly counterintuitive play, before an inflatable penis can rise to its full potential, first, it's got to relax. That's what causes normally constricted arteries to open up (dilate), allowing blood to rush in and fill the porous tissue, turning a squishy, flaccid sausage into a hardened ramrod. To withstand such increased pressure without bursting, the inflatable penis relies on the same basic engineering as a puffer when it blows up: lots of alternating strands of stretchable collagen.

In contrast, some mammals have actual bone in their boners. Called a baculum, an internal penis bone, or *os penis,* this appendage can be found in rats, cats, dogs, bears, and even our closest relatives, the chimpanzees. Females of these species have a complementary (and flowery-sounding) baubellum, or *os clitoris,* which is a rare case of Nature being fairer in sexual handouts. Both of these bones help to erect sensitive tissue, which likely makes for easier sexual stimulation and reception in both sexes.

Seals, sea lions, walrus, and polar bears took this relic of their terrestrial heritage with them when they returned to the sea, and proudly sport some of the sturdiest male erections around. Walrus have the longest of these penis bones on the planet, at nearly two feet in length. They are slender, ivory-colored rods with a slight curve, giving them the appearance of an elephant's tusk. To promote a svelte form, a male walrus, like most marine mammals, retracts his penis until the final moment before actual penetration.

Finally, whales and dolphins (along with cattle and pigs) have penises that gain support from a thick cord of tissue that keeps their members perpetually at half-mast. Such a fibro-elastic penis provides a combination of firmness and flexibility, and makes for near immediate erections. This elasticity also allows for a copulatory structure that is both bendy and long—a helpful feature for the friskiest of dolphins and most giant of whales (collectively known as cetaceans).

Famous for their impressive libido, several species of dolphin engage in sex for sport as much as for reproduction. It may be for fun, or to establish social bonds, or in some cases to assert hierarchies. For example, juvenile males will have sex with other males, forming strong ties, or they may use sex to establish dominance. These same males later in life will often form gangs and cooperatively corral females and mate with them—sometimes quite forcibly.

In a momentary disruption of their streamlined form, horny male dolphins protrude what looks like a pointed cow tongue from their genital slits. They use this stout saber to explore one another's bodies, choosing to enter and exit all kinds of available openings in all sorts of available mates. Male-male sex, via genital slit as well as the anus, has been documented, as has sex between closely related species of dolphin.

Libertines of the underwater realm, dolphins don't mind a little variation in sexual position: horizontal and vertical penetrations are both common. They may do it midwater or at the surface; the frolicking pair may slow down or continue to beat their tails with gusto, making for some high-speed sex thrills. Ejaculation is as swift as their erection: actual penetration need last only a few seconds. A male dolphin's rapid response rate also extends to recovery time—he's ready to go again in only minutes and can do so repeatedly for over half an hour.

A similar freedom of phallus expression occurs in the larger whales. Dr. Phillip Clapham of NOAA first started studying cetaceans off of Cape Cod, where he used to lead whale-watching tours and got his first glimpses of male members in action. During the springtime migration up the coast, North Atlantic right whales gather off the Cape and can get a little frisky. One time, Clapham recalls, "there I am standing on the upper deck, calling out where to look and talking about the natural history of the whales when a male and female start doing it right next to the boat." Belly to belly and just below the surface, the male's long pink shaft was definitely visible upon the approach.

Below Clapham on the lower deck were a woman and her five-year-old son, a mere few feet above the copulating couple.

"The kid turns to his mother and says, 'Mommy, what are they doing?' Without hesitation, the mother replies, 'They're saving the whales, dear!' It was brilliant."

But this is not even Clapham's favorite story about whale sex. That one, he says, belongs to a friend and colleague, Dr. Bruce Mate, director of the Marine Mammal Institute at Oregon State University. Like right whales, gray whales provide for some rather X-rated whale watching. Mate notes, "In Mexico, the tourists are always watching for 'Pink Floyd' and are frequently 'rewarded' with sightings."

Tracking down a female gray whale on the west coast of Baja, Mate was approaching a courting triad (two males and a female), whose sexes were obvious. As Mate went to tag one of the whales, the female swam right under his boat, chased by a male. Apparently, the boat seemed like good cover to her. The male, however, didn't take the hint. Instead, swimming up alongside her, he rolled and then swung his enormous pink penis right into the researchers' inflatable boat. "This huge phallus prodded around, trying to find the female's genital opening," says Clapham, laughing. "Mate says he had this moment of realization . . . he was sitting in the world's largest diaphragm." In recounting the tale, Mate admits to "jumping around to dodge 'contact.' The thought of a camera never crossed my mind, but I would have given $1,000 for a photo or video of the event."

Even Sir David Attenborough hasn't gotten *that* up close and personal with wildlife.

Although whales and dolphins engage in lots and lots of sex, they do not make lots of babies. Twins are extremely rare in cetaceans, with females usually producing one offspring at a time, often spaced over several years. Right whale females breed every three to five years; sperm whale females may give birth even less frequently.

Sex by way of a penis has its drawbacks, especially for species where females carry developing young. There is only so much room in the womb. Pregnancy is taxing for females, necessitating breaks in between birthing babes (plus, in species such as sperm whales, young may nurse for several years before becoming fully independent). From whales to walrus to sharks, this low level of output means these animals face greater risk of population collapse when hunted. In cetaceans, a low level of reproduction combined with a complex social system in some species makes them extremely vulnerable to fishing—as well as to other impacts such as pollutants—and can make it hard for some species to bounce back after they have declined.

When it comes time to penetrate to procreate, though, reaching the female's opening is just the start of the penis-sperm journey. The internal female anatomy of several whales and dolphins presents a complex of pathways and tunnels, as we'll explore in more detail soon. For these species, successful fertilization depends on a penis and sperm with the capacity to travel the twists and turns of the female vagina. A little length never hurts in such situations, but having enormous volumes of sperm also can help. Those of the North Atlantic right whale top off at about a half-ton each, the largest testes on the planet. But they aren't the only male cetaceans to have enormous packages. In the smaller, approximately human-sized harbor porpoises, the testes are a whopping 4 to 6 percent of the total body weight, or roughly eight pounds. That's more than one hundred times greater than what the average human male devotes to his testicular heft. As Nature is all about efficiencies, there's got to be a reason for carting around that much sperm-making gear, and it's likely the presence of some stiff competition with other males.

Although little is known about mating behavior in many cetacean species, testes size is likely related to the level of sperm competition, as it is in other animals. Scientists hypothesize that those species with the biggest sperm factories are probably facing lots of competition from other males—in other words, there is lots of sex going on and males can't really control access to females.

On the other hand, in species where a few males win access to and monopolize the females, the dominant males do not need a lot of sperm to outcompete other males—what they need is good fighting gear. Recent studies show that there appears to be a trade-off between investment in the adaptations for winning fights before sex—such as enormous tusks, battle teeth, or a big body size—and traits that aid with winning battles after sex, namely, gigantic testes to pump out loads of sperm that duke it out in the female's reproductive tract. So it is that one law of Nature seems to be the bigger the brawn, the smaller the tackle.

THE LONG AND SHORT OF IT

On one end of the oceanic penis length spectrum lies (or slightly protrudes) the genital papilla of some species of fish, such as sculpin. More of a pimple than a penis, this tiny, barely there bit suffices to deposit sperm inside the female. To assist with anchoring into the female, the male hooks an elongate bony anal fin around the female and holds her against the nub in order for sperm transfer to happen. At the other end of member measurements, however, is the blue whale's behemoth, which can be up to twelve feet long—the longest penis on Earth.

No surprise, extremely long phalluses tend to draw attention. There is a rich dialogue and debate in both scientific and popular literature about the pros and cons of a lengthy male unit—mostly in regard to how they are received by females. To be sure, female choice is a strong selective force on the length of the male penis (as it is on its overall shape). But, in contrast to women's magazines' extensive explorations of the relationship between penis length and female pleasure, in nature, female preference for penis size relates to screening for good genes.

To females, giant genitals (like a deep voice) can be a sign of fitness. Just as a peacock's investment in lavish, iridescent tail feathers indicates strength and health, so too can a mighty member indicate a strong and virile male. Over generations, this kind of female selection helps drive longer and longer male members throughout a population.

In the race to reach an egg, the closer a male's sperm can get to the finish line, the greater the chance of success. So, longer shafts can be advantageous to males for this reason too. Especially in species where multiple males mate with a single female, a longer phallus that can reach beyond the sperm drop-off zones of previous males gives its own cargo a head start. It may also help push aside competing sperm.

Left unchecked, however, that combination of selective forces would lead to males with enormous and ultimately incapacitating penises. Giant male members can be hazardous. Remember the well-endowed mosquito-fish? Reduced mobility and reduced functionality put the brakes on ever-extending penises.

That's not to say some enormous phalluses don't exist. But to truly appreciate their scale we need to consider length related to the size of the owner. Looking at this ratio evens the playing field between, say, barnacles and blue whales, allowing for far more colorful comparisons. For example, a one-hundred-foot blue whale with a ten-foot penis has a penis-to-body ratio of about 1:10. People, on the other hand, come in at about 1:13, assuming average human penis length (erect or stretched out flaccid) is about five to six inches and assuming a six-foot-tall male. So blue whales have us beat, but not by much. Move into invertebrate territory, however, and it's a whole different ball game.

As noted earlier, the ostracod has an intromittent organ up to one-third its body length, a ratio of 1:3. With bodies encased by hard shells, male ostracods have to have some length in order to navigate around all that armor. The male queen conch's verge is a similar scenario, approaching nearly half the total length of its body. But that's nothing compared to the world-record holder. If there were a penis Olympics, the following species would win the decathlon, sporting not only the world's longest proportional penis, but also a remarkably agile and malleable one.

To witness the world's arguably most athletic penis in action requires little more than a visit to a tide pool. Once there, focus not on the darting fish, or the scurrying crab, but look to the small white protrusions that encrust the rocks at the high tide line. These sharp, jagged peaks house within them a small crustacean that lives its entire adult life lying casually on its back, waving its enormous penis up into the water and periodically snaking it across the tide pools. You may know it as a familiar adornment to pier pilings, rocky shores, and even humpback whale heads. Say hello to the barnacle—owner of the largest penis (relative to body size) on Earth.

Don't let their almost inanimate appearance fool you: encased inside that tough outer shell is a horny crustacean that goes the distance for a little nookie. Look closely and you may catch sight as one slowly pokes a tapered, transparent tube out of the central crack in its shell. Then behold as this firm phallus unfurls up to eight times the length of the barnacle itself.

Even Charles Darwin couldn't stifle surprise and excitement over the barnacle's gigantic male proportion, writing, "The prosciformed penis is wonderfully developed."

Relatives of shrimp and crabs, barnacles are distant cousins of the well-endowed ostracod, but barnacles have given up the mobile life. After a few weeks floating in the water as larvae, they settle down onto a hard substrate—be it seafloor or turtle shell—and never move again. They glue themselves to the bottom in a headstand position, with legs sticking up, and surround themselves with hard overlapping plates. When the water is flowing, they peddle their feet to pull in small particles of food; at low tide, they can close a hatch, keeping the moisture locked in until the waters rise again.

The real kicker for barnacles is that despite being stuck where they land, all but one species (the gooseneck barnacle) still require internal fertilization. Although most are hermaphrodites, they don't self-fertilize. Instead, they rely on an impressively expandable penis to prod and poke their way into the domed homes of their nearest, and at times not so near, neighbors.

Like the long sweeping pipes of a farm's giant irrigation system, the barnacle swings its phallus in wide arcs to fertilize the encrusted shore. The longer the pipe, the greater the area of irrigation. But in a watery world, unlike on a farm, a longer penis also faces the problem of greater drag. Even the calmest shorelines at times are swept by fierce currents that could bend and distort long love shafts. Similarly, the roughest coasts may have their quiet seasons, when even more extended members would be welcomed. But this is no problem for barnacles, which turn out to be remarkably adept penis shape-shifters.

As hard as they are, as rigid in their locales, barnacles have remained flexible when it comes to precisely proportioning their penis in response to wave action. In experiments, barnacles transplanted from calm to rough environments adapted shorter, stouter units better able to resist strong flows. Barnacles from the rough neighborhoods, in contrast, expanded their chunky sticks into elegantly long, wandering tendrils once in calmer waters. The transformation is not necessarily immediate—these changes were measured after about five months. Still, these studies show that barnacles not only impressively expand their phalluses, but may morph their shapes at will.

OUTSIDE THE BOX

Four attractive women sit sipping cocktails at one of the most popular cafés in the city. Beautiful, young, and carefree, they enjoy the late afternoon sunshine. Weaving through the outdoor patio, a young man quickly wends his way to the bar, passing by the ladies' table. In his rush, he accidentally bumps the chair of one of the women. He apologizes, steadying the chair and bowing slightly as he backs away toward the bar. Only later, when she's back at home, does she discover this was no accident at all. There, stuck to the back of her sweater, glistens a small, delicate packet full of sperm. This makes the fourth time she's been tagged in as many months. She sighs as she tries to picture the man again. Recalling that he was fairly good-looking, she scrapes the sperm pack off and tucks it away in her personalized sperm bank for safekeeping.

    It's always good to have options when it comes time to reproduce.

——— ——

IN THE WORLD of squid copulation, sex can be like a pat on the back. At least, this is the approach in shallow-living species. Deep-sea squid are a different story, and we'll get to them soon. For most squid, the kind that make up our calamari, for instance, a male relies on his modified fourth left arm, the hectocotylus (pronounced *hecto-cot-illus*), to deliver the goods. This specialized arm can look very different across species; in common, however, is the absence of normal suckers at the tip, which are often replaced by small fleshy flaps called lamellae that hold and transfer the spermatophores.

In octopus, which also have hectocotyli, the tip is made of erectile tissue that can stiffen and aid in insertion up the female's funnel. Rather than slapping a sperm bundle onto the outside of the female, as often happens in squid, octopus snake their arm-penis—their third right arm—into the female's siphon and up into her oviduct. Octopus sex takes some maneuvering that demands more extended contact between males and females. This doesn't always work out so well for the males. Female octopus have been known to strangle their sperm donors. This may be the reason why many males mate at extremely extended arms-distance from the female, attempting to avoid her clutches, or sneak-attack from behind, jumping onto her head to remain out of reach.

In contrast, when a large, aroused male squid sees a female, he is likely to flash a flirtatious color pattern across his body and then mate with her in one of two positions: head to head or, as has been observed in common squid, in parallel. When courting in this latter position, the male slips his fourth right arm up into her funnel to hold her in position and then uses his fourth left arm to press a sperm pack onto her mantle—the thick sheath that covers the body and protects the organs of squid, leaving the head and tentacles to stick out the end. In other species, the sperm packets wind up around her head, neck, or even a special receptacle near her mouth. Pack placement is species-specific and can be highly variable, which goes to show that in some squid, males may be rather opportunistic in making their deposits. The transfer itself is swift, happening in the blink of an eye (I'm guessing the females prefer it this way, as having a large male arm stuffed up one's siphon isn't likely to be all that comfortable). A female can then use this sperm to fertilize her eggs as she extrudes them.

Observations of the commonly fished European squid, however, reveal that sneaker male squid, like their cuttlefish counterparts, skip all the pleasantries and instead dive-bomb females, seemingly at the moment when they are carrying a recently extruded egg capsule. When a female with the visible bulge of an egg capsule in her arms is within striking distance, a sneaker male appears to nonchalantly reach up inside his own body, grab a sucker-full of spermatophores, and attempt to slot them right between her outstretched arms, apparently aiming for the egg capsule. Relying on superb swimming skills and agility, these sneaker males attempt to mate regardless of whether a female is alone, guarded by a much larger male, or even mid-act with a partner. Deftly as a pickpocket, these males leave rather than take the valuable wares.

Just imagine those bar scenes. Males subtly slip little bundles of potent DNA packs onto the back of a jacket or skirt hem. For females, the risk radius for being mated would be a few feet, making every male that walked by a high-level threat.

In contrast, some deep-sea squid don't have a hectocotylus. For years, the question of how these giant predators managed to transfer sperm remained a mystery. Scientists theorized that the squid's terminal organ—a term that refers to the elongate tube that attaches to the sperm storage sac and is not a morbid reference to the fact that many squid die after mating—must play a role in getting the sperm to the female. But they weren't sure how it happened. One thought was that, perhaps, if the male could hold the female head to head, as observed in other squid species, the terminal organ—aka squid penis—might be able to reach her.

A few intriguing lines of evidence pointed in this direction. First, in some species, the terminal organ was longer than the mantle, providing evidence that it could extend past the male's body to reach a female. Second, remains inside of a sperm whale stomach included interlocking beaks of a male and a female giant hooked squid, a deep-sea squid lacking a hectocotylus. This finding supported the head-to-head mating hypothesis (and indicated that this sperm whale probably scarfed down the couple mid-copulation. Sex can be a dangerous distraction). But since nobody had ever seen deep-sea squid getting down to business, it was impossible to know for sure if this squid penis could indeed get the job done.

Then, in July 2006, scientists got their first glimpse of a terminal organ demonstrating its full potential.

Off the coast of the Falkland Islands, a giant hooked squid was brought up from three thousand feet during a scientific research expedition. The change in pressure as it rose from its deep-sea habitat had nearly killed the creature, and it lay on the deck of the ship, tentacles quivering. With the squid well on its way to death, the scientists on board cut open its mantle to study the innards. As the researchers peeled back this outer casing, they saw the male-defining, pale white terminal organ lying inside, just long enough to reach the edge of the mantle. So the scientists noted the animal's sex in their notebooks, and for a brief moment, everything was as it should be.

And then, right before their very eyes, that mighty member began to elongate and stiffen. It extended past the mantle opening, past the head and giant eye, all the way to the very tips of the outstretched arms. It was the equivalent of a man reaching his arms overhead and inflating his penis to his fingertips (with his erection starting at the knees).

The dying squid, in a final testament to its virility had boasted a mighty and last erection. The total length of the squid's mantle was just over one foot; the length of the fully extended terminal organ over two feet—more than enough to reach a female during sex and second only to the world-record-holding barnacle for lengthiest penis relative to body size.

That same year, another discovery solidified the respect we owe to the giant, inflatable penises of deep-sea squid.

In 2006 and again in 2012, remotely operated vehicles (ROVs) captured on film two squid midcoitus and unbashed enough to continue their exploits despite the bright lights and snapping cameras. When Dr. Henk-Jan Hoving with the Helmholtz Centre for Ocean Research Kiel and Michael Vecchione with the Smithsonian analyzed the footage, they found themselves witness to quite a demonstration. The mating squid were in a 69 position with the male on top, but inverted, lying on his back, with his belly side up. It was the first time researchers ever saw the long terminal organ in action. And what "action" it was.

To truly appreciate the acrobatics involved, picture a man lying on his back, on top of a woman who is also lying on her back, their heads in opposite directions. He then sports an erection that sends his penis arcing up and over his shoulder to wrap around his back and insert into the woman lying beneath. We are talking about a penis equal in length to the male's body doing a reverse-backflip implantation maneuver.

In the video footage, the male's terminal organ reflects the ROV's lights, looking like a glow stick arching out from beneath the deep purple mantle. The male grasps the female in a triple-clasp, three sets of arms holding her tight. Despite the snug hug, the equally velvety and equally purple female continues to swim languidly along, not seeming all that impressed by her mate's reverse-inversion insertion.

Bearing witness to a deep-sea squid erection in action is remarkable for two reasons. First, it is admittedly one hell of a long penis. It ranks second in the animal kingdom in terms of proportional length, and is the longest of any mobile invertebrate on the planet. Second, seeing the reach of this penis helped begin to answer how squid lacking a hectocotylus could transfer sperm from the male to the female. Or, in some cases, male to male.

From studying preserved specimens, Hoving had seen males often covered in implanted sperm packets, but he didn't know if these were from another male or just self-inseminations. This can happen when squid are captured and hauled up from the depths: the males ejaculate and their sperm packets attach to anything nearby. To figure out the riddle, Hoving and his team decided to comb through hundreds of hours of video footage from past ROV missions. Scanning the images, they could see that living males were covered in implanted sperm packs as much as females were (conveniently, the attached spermatophores are white in appearance, making them relatively easy to identify against the dark-colored squid). Amorous squid do not appear all that discerning when it comes to choosing a mate.

In the pitch-black waters of the deep, where opportunities for mating are few and far between and where it can be difficult to distinguish between male and female, males would rather waste a few sperm packs inseminating another male than miss out on a rare mating opportunity by mistaking a female for a male. Might as well hit up everyone, just in case.

Using the same creative approach, comparing museum specimens with live observations, also helped Hoving and Vecchione confirm sperm placement in the acrobatic mating pairs caught on film. A museum specimen of a female from the same deep-water species had sperm bundles deeply implanted in her mantle near where her fins join. This is right where the videos showed the male's terminal organ seemingly contacting the female. The final step in the journey of squid sex, however, remains a mystery: we have no idea how the sperm actually make it to the eggs. Hoving has made some interesting discoveries that start to illuminate the process, including that squid sperm packets have a mind of their own. When the spermatophore extrudes through the terminal organ, the tip of the penis acts like a trigger to spring open the packet, freeing the sperm bundle—known as a spermatangia—to wiggle and burrow under the female's flesh. The spermatangia contain millions of sperm cells that must now somehow make their way from under the female's skin to the site of fertilization. In female squid with sperm storage pouches, the sperm migrate or are moved from the site of implantation on the body to this sperm receptacle. In other species, males release sperm packets directly into the mantle cavity where sperm can fertilize eggs on their way out of the oviducts. But sperm deposited on the tail or embedded in the back of the female's head have a long way to travel to reach the oviduct or even the opening where eggs might be extruded.

This is the next mystery, still kept by these "supermollusks" of the sea, as Hoving refers to them. And super they are. As a group, squid can fly out of the water for short distances; their camouflage capabilities exceed that of any special ops; they have complex eyes with excellent vision; they can produce their own light; and they can extend their penis to a size equal in length to their bodies. Not even Superman can do that.

But as impressive as squid may be in terms of lengthy male members, there are a few penis tricks that other species have perfected that leave squid looking merely average.

DIS-MEMBER-MENT

While flatworms battle with their lithe lances in close combat, and squid sport impressively extendable equipment to slap some sperm on a female, other oceanic males turn their penises into projectile mating machines.

An unusual type of octopus, the argonaut, shirks the bottom-dwelling life of most of its relatives to instead roam the open sea. The females can often be found extending elegant arms from within a beautiful, paper-thin shell, from which their other common name derives: paper nautilus. A translucent, pearly white and shaped like two abalone shells hinged at the outer edge, this fragile secretion serves as a chamber where the female will brood her fertilized eggs. Though her body is not attached to the shell like an oyster or conch, argonaut females will curl up inside their delicate mobile homes, using trapped air bubbles to help regulate their buoyancy as they roam the midwater realm.

Growing to up to about a foot and a half in length, the female argonaut is not a large octopus, yet she dwarfs her male counterpart. Coming in between an inch and an inch and a half when fully grown, a male argonaut seems less than intimidating. But any female who wakes up the next morning to find her gills stuffed with a still-wiggling dismembered penis might feel differently about that.

Yes, the male argonaut can detach his hectocotylus and send it to inseminate a female from afar. Though males may be small in stature, their penis-arm is impressive in size—longer than the male's own body. Until mating, he stores his specialized appendage in a pouch located beneath his left eye. When opportunity strikes—a once-in-a-lifetime affair for the male—he unfurls his love limb from the thin case and then, in an extreme example of Zen-like nonattachment, he amputates the precious appendage and swims away. His hectocotylus must fend for itself in a final wormlike sprint to the female. The male dies soon after; the female lives on to mate again, severed penis firmly implanted on the inside of her mantle. Females

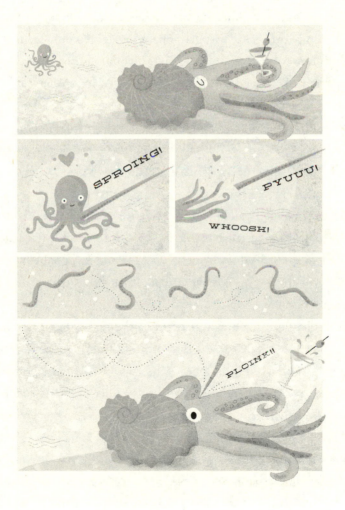

have been found to contain multiple hectocotyli from different males, all shoved up their mantles, indicating that a single female argonaut may be able to brood eggs fertilized by the detached penises of multiple partners all at once. But we really can't be sure. The sex lives of these open ocean octopus remain for the most part an enigma. And they've fooled us in the past.

When first discovered, the dismembered members of the argonaut male were thought to be a species unto their own—some kind of strange parasitic worm that latched onto the females. They were even given their own genus, *Hectocotylus*—the name stuck, and it is now applied to the specialized penis-arms of all cephalopods. How exactly the male manages to turn his penis into a projectile, or why he goes to such extremes, remains unknown, in part because living argonaut males are rarely seen.

We do know, however, that the male argonaut is not the only one who finds a detachable appendage a thing of beauty. Many species can drop and regrow a limb in the name of survival: sea stars can easily re-arm, or lizards regenerate a tail. But all these are mere parlor tricks when compared with regrowing a penis. And I'm not talking here about regenerating a damaged penis after some kind of misfortune, as the conch does. One species can make entirely new sexual units, sometimes every year.

In the segmented palolo worm, a relative of earthworms that lives in the sea, the full moon triggers a mass sex fest whereby the back ends of males and females swell with sperm or eggs and then split off from the main body of the worm. Known as epitoky, it's the self-sacrifice of one's rear end in the name of sex.

While the head remains burrowed in the seafloor, the dismembered unit rises in slow spirals toward the surface, like an escape pod from the mother ship. Though it lacks a head, the epitoke (the body part undergoing the epitoky) comes endowed with its own set of eyes—one per body segment—with which to orient toward the heavens. Upon reaching the surface, the budded backside breaks open and spills its gamete cargo into the water to mix with the spewed sperm and eggs from thousands of other severed worm butts that have amassed at the surface. Meanwhile, down on the seafloor, the front end of the worm begins to regenerate a new backside.

The regrowing process typically takes several months and repeats over the course of the worm's many years (some even live decades).

In contrast, the ragworm, from the family Nereididae, doesn't cut off its sex parts; it *becomes* them. Fishers may be familiar with these worms as bait, or aquarists may recognize them as food for their tank fish. In Vietnam they are even eaten by humans as a delicacy. In the shallow coastal zones and estuaries of the world, they are a common and important food source for birds and fish. But it's their transformational sex life that really sets them apart. As they approach maturity, these worms undergo a dramatic metamorphosis. It's another form of epitoky, but this time there is no budding backside; instead, the entire worm becomes the sex organ—a giant swimming sperm or egg sac.

To transition from bottom-crawler to surface-spawner, the ragworm's hundreds of bristlelike crawling legs morph into paddle-shaped swimming appendages; its gut atrophies as its body fills with sperm or eggs, and its eyes enlarge in their now more muscular swimming front ends. Once transformed, the ragworms await the full moon. The light from the glowing orb triggers the ragworms to lift off the seafloor en masse, swim to the surface, and burst open in an explosion of sperm and eggs. It's a fatal, but effective, mass orgy and a rather tragic version of puberty.

The only animal known to have disposable-regrowable penises is *Chromodoris reticulata*, a tropical nudibranch found in coastal waters of the Indo Pacific. Not only are these sea slugs badass enough to live through self-amputation of their man parts, they also produce a backup penis that's ready to go within a day. These beautiful, bright red-and-white dappled sea slugs are basically a penis PEZ dispenser, harboring a long, coiled structure that unfurls penis after penis after penis. This tightly spiraled mass appears to have length enough to allow for at least three sex bouts, given some rest in between.

As reciprocal mating hermaphrodites, two slugs tightly coordinate their mutual copulation. First, they approach and touch each other's genital openings as if to orient themselves. Then, the pair separates and turns to face opposite directions. They line up along their right sides, make a few small adjustments to their partner's protruding penis, and then commence and complete the simultaneous insertion. Minutes pass. Insemination occurs. Then, each slug begins to pull away from the other, the two penes stretching like elastic bands between the parting bodies. They stretch and stretch, thinning as they are pulled to impossible lengths. Eventually, one slug tugs its penis free from the other; the second slug soon follows. They crawl away, their elongated tackle dragging around behind them like a woefully overstretched slinky. A few minutes later, they simply drop this extra baggage and begin to grow a new one.

Each slug mates with the other, exchanges sperm, and then parts with its own penis. It's all quite civilized, really . . . until you look under a microscope. Then, the sticky nature of these dual penetrations comes into focus.

Row upon row of backward-pointing sharp spikes cover the head of the penis. These spines act like Velcro, ensnaring sperm and dragging it out of the genital opening upon disengaging. While it is possible that these hooks grab some of the depositor's own sperm, a more likely scenario is that these thorny heads scrape out sperm left by rival males—of which there are many.

These bottom-gliding beauties live as cozy neighbors among the corals. An abundance of mating opportunities each season results in strong sperm competition among the "male" half of the population. A long phallus with which to embed deeply and a swollen, thistle-like head with which to scrub out previously deposited sperm can be very advantageous under these conditions.

Multiple penises are actually quite common in the sea, with some species sporting them all at once.

DOUBLY ENDOWED

Ejaculation—it's a mighty force of nature. But for male sharks, it's more of a slow leak than a fire hose. So they have to supplement their less-than-impressive eruptions with some hydraulic pump action, turning their penis into something akin to a Super Soaker water gun. Because sharks have two intromittent organs, they've got this additional built-in firepower on each side.

Superficially, shark penises, called claspers, look like a mammalian penis, but they are actually extensions of the pelvic fins with a simple groove down the middle. These elongations, made of calcified cartilage, become ever more rigid as a male matures.

Near the base of the tail on the underside of the shark, this matching set of sausage-shaped rods look exactly like what they are, yet scientists originally thought males used them to grasp the female, thus the name. In reality, claspers don't clasp at all; they poke and then anchor like a grappling hook inside the female.

A pair of claspers is evident from the moment a male shark is born. But a newborn male shark pup has floppy claspers. As the young lads mature, the flaps elongate, curl inward, and harden into what look like two giant cigars. Some species of shark can keep their fully formed tackle tucked up flush against their undersides, reducing drag. Even so, there is no hiding their dual dongs. Next time you're standing beneath one of those glass tunnels at a fancy aquarium, check out the underside of the shark or ray cruising above you. The males will be obvious.

That type of identification is not something you can do easily with a goldfish, by the way. They, along with species from cod to snapper to salmon, keep their sex far more under wraps. But sharks are a different kind of fish: they are Chondrichthyes—cartilaginous fishes, named for the stuff of their skeletons, the same material that builds our noses and ears (as opposed to the calcified bone found in our bodies and those of bony fish). Ancient chondrichthyans split from bony fish about four hundred million years ago, evolving into modern-day sharks, skates, rays, and chimeras. And they brought with them some impressive sexual outer-ware.

A ten-foot-long great white shark, for example, might sport a pair of claspers with each reaching about three feet in length. Some of the greatest porn shots of the animal kingdom are of the enormous claspers of leaping great whites, backlit by a setting sun as the shark hurls its body out of the sea to snatch up a fleeing seal.

Studying shark reproduction is a challenging specialty: extreme diversity in addition to the highly elusive nature of these animals is a perfect recipe for continual mystery. Including the skates and rays, there are more than 1,200 species of cartilaginous fish in the sea. Nearly half of them live in waters deeper than 650 feet. We know hardly anything about most of them, least of all their sex lives. What we do know is that they all have internal fertilization, requiring close contact between males and females and the effective deployment of the clasper.

Dr. Dean Grubbs of Florida State University has been studying shark reproduction for twenty years. He notes that when a male pushes his clasper into the female's cloaca (the general opening that serves as entry and exit for penis, pee, poop, and, in many species, the birth of live pups), it flares open and kind of folds back on itself, like a hand opening up. Some claspers are adorned with what Grubbs refers to as "mild anchoring systems," while others have pointed, hook-like barbs at the end—not so mild. (In at least one species of shark, the blue shark, females have particularly thick-walled vaginas, which may have evolved in response to evolution of these pointed appendages.) The anchors help keep the male docked inside the female long enough for him to unload his cargo.

Because claspers are just curled extensions of the pelvic fins, there is no tube (such as a urethra) that connects the sperm storage site down the length of the "shaft" to the tip. Instead, the urogenital papilla—which looks like a nipple but marks the opening through which the sperm come out—is up in the male's cloaca near the base of the clasper. So, unlike in mammals, where males ejaculate through the end of the penis, when sharks ejaculate, the sperm just kind of oozes out and would travel down the claspers in slow motion—if not for the super-soaker action.

Near the base of the clasper sits a tiny pore, just next to the genital papilla. There is one pore by each clasper, and each pore connects with a long, internal, slender sac that runs the length of the shark's belly. In immature sharks, the sac is only partway up the torso, but in mature males, it reaches all the way to the gills.

"When a mature male decides it's time to get busy," says Grubbs, "they use muscular contractions to create a vacuum that effectively sucks the seawater in through the pore. The sac fills up like a water balloon." To penetrate the female, a male often uses the clasper on the far side, crossing it under his own body and then pressing it into the female. This means that if the male is on the female's left, he will use his left clasper to inseminate, and here's why: as the clasper pivots across the body to insert, the small pore comes in direct contact with the groove on the clasper. As the male releases sperm through the genital papilla, he contracts his muscles around the internal sac, squeezing water out of the pore. The squeezing creates a high-powered jet of seawater to flush the sperm down the clasper and into the female. It's a remarkable and unique ejaculatory system.

One of the few cases where shark sex has been directly (and relatively unobtrusively) observed is in nurse sharks, whose tropical shallow water habitat offers the chance for an extended viewing. Nurse shark sex looks like a giant coppery corkscrew of two (or more) tangled, twisted bodies, with the male biting onto one of the two large pectoral fins of the female and wrapping his body around hers. The technique leaves battle scars. Similar wounds on females from blue sharks to sand tigers indicate this is likely a widespread mating habit, used by males as a way to maneuver females into position.

With a firm grip on her winglike pectoral fin, the male nurse shark uses his tail as a lever to turn the female—sideways, onto her back, or even vertically head-down—to provide access to her underside. In female sharks, skates, and rays, the cloaca is located just between the two pelvic fins, same as it is in males. Once in position, an achievement often facilitated by the presence of a second male who helps keep the female in place, the first male maneuvers the nearest clasper to penetrate the female. Having two penises has its perks.

Wrestling an unwilling female shark and flipping her into position is no easy task for the male. It's clear that at first she wants no part in the event, and who can blame her? But in a strange, S&M kind of twist, such courtship biting seems to subdue the female and even entice her to acquiesce. In some species, where love bites are extreme, females have extra padding to reduce the impact of the bites. But as females thicken their hides, males sharpen their teeth. In small shark species and many rays, including stingrays, males have specialized teeth designed just for mating purposes—we know these teeth are specifically for sexual activity, as opposed to eating teeth, because the teeth appear only during mating season. Sex teeth to enhance male shark sexual performance—sounds like a vampire's version of Viagra.

Where witnessed, shark sex appears to be a rather rushed affair. It makes sense, what with the male nearly suffocating on the giant pectoral fin crammed in his mouth, and the female eager to shake the male off. Thus far, male sharks appear to use one clasper at a time and disengage after insertion for about twenty to thirty seconds, though sometimes it can be longer. In nurse sharks, at least, as one male finishes, another is often waiting in the wings, and one female may mate with a group of males within quick succession. Whether this is by force, or by choice, is difficult to know.

Sharks are not the only species to sport double-hung natures. As we've already explored, ostracods have two penises, as do copepods. Other species with multiple penes include blue crabs, which use a pair of pleopods to implant sperm packs inside the paired genital openings of the female, and sand-dwelling *Pisione* worms, one of the few polychaetes with internal fertilization. These worms have multiple penises or vaginas, one along each body segment for over ten segments in some species—when they have sex, it's like a giant zipper zipping up.

FOR SPECIES RELIANT on some form of sex by penis, the plethora and diversity of penises in the sea is only half of the mating equation. Successful sex can only be achieved through appropriate docking and delivery of sperm into some kind of female receptacle—which turn out to have their own colorful spectrum of size, shape, form, and function.

— 5 —

INNER CHAMBERS

Influencing Sex from the Inside Out

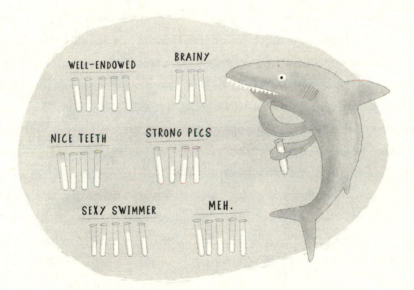

WELL-ENDOWED   BRAINY

NICE TEETH   STRONG PECS

SEXY SWIMMER   MEH.

SEX-SEA TRIVIA

- *A whale vagina can be so convoluted, sperm need a good GPS.*
- *Some sharks can become pregnant nearly four years after they last had sex.*

- *In fish and worms, dwarf males make great sex slaves.*
- *Virgin birth is alive and well beneath the waves.*

SEX-SEA SOUNDTRACK

1. "Maneater"—Hall and Oates
2. "Billie Jean"—Michael Jackson
3. "Like A Virgin"—Madonna

"WHEN IT COMES DOWN TO IT, FEMALES HAVE THE home field advantage."

This is how Dr. Sarah Mesnick at NOAA's Southwest Fisheries Science Center sums up the situation for internal fertilizers. Males invest tremendous resources in growing big, developing structures for fighting other males or flirting with females. They invest in enormous penises and in some cases giant testes. All this effort is aimed at closing the gap between sperm and egg. But ultimately, that final stretch—be it millimeters or feet—is played out inside the female's reproductive tract.

And it can be quite an obstacle course.

The same evolutionary pressures that drive development of a male's efficient and effective genitalia similarly drive the evolution of the reproductive parts of females. The form and function of her sexual apparatus is equally focused on securing the destiny of her genes. This means both male and female battle for ultimate control over the paternity of her offspring, and that has led to some remarkable female adaptations.

Unfortunately, the science world has traditionally studied the evolutionary biology of male genitalia more than female genitalia. While it is generally easier to observe and study penises than vaginas, this imbalance has more to do with a long-standing assumption: namely, that the male sex is in the driver's seat when it comes to evolution.

This kind of bias was the norm back in the Victorian era, when scientists assumed females—and their lady parts—were passive participants in the game. The overarching view was that males battled one another for access to females, and he with the most-fit genes won the prize: successful reproduction. The winner's genes spread, influencing the fate of the species. Darwin first challenged this view when he noted that females could influence how traits evolved by choosing the male with the brightest colored feathers or longest, flashiest tail. Such sexual selection could lead to development of traits that were not about survival, but about scoring mates.

Today we know there is much more to female choice than meets the eye. Cryptic female choice—the ability for females to influence whose genes win out, during or even after copulation—is a major force of nature. Through behavioral, morphological (shape), and chemical means, the female body has developed different ways to make certain copulations more or less successful, and in some cases, developing offspring more or less viable. In many species, females can store sperm, and through physical manipulation, hormones, or other mechanisms, regulate which sperm are released into the reproductive tract. The vagina itself can also be a screening device. Secretions may preferentially wipe out one male's sperm over another's based on genetic compatibility, or, as we are about to see, the shape of the vagina itself may be a selective force. From the sperm's perspective, the female reproductive tract represents a battlefield not only where competition with other males' sperm may ensue, but the field itself is beset with obstacles designed specifically to eliminate some while attracting others.

That is why it is surprising that this preference for penis studies over vagina research is still on the rise. And it is troubling for two reasons. First, a bias in research subjects skews our basic understanding of why species evolve and has likely underestimated the role of female choice and selection in influencing the course of evolution.

Second, on a practical level, this genital gender bias prevents a basic understanding of reproductive biology, without which we can't effectively manage natural populations. If the goal of fisheries management is to optimize the growth potential of wild populations, we've got to know the basics about what makes a population grow and shrink.

Studies where male and female genitalia have both been examined have proved insightful in several regards. For example, in some species of ducks, females are constantly pressed into "forced copulation" (the scientific term for rape). In these species, sperm competition can be strong. While males have an elaborate and long, counterclockwise-curving penis, females have a complex, *clockwise*-spiraling vagina, which proves a formidable defense against fertilization. Despite the multiple unwanted mating events, females appear to retain control of fertilization to a significant degree. This, in turn, has major implications for how many (and which) males are actually contributing to the gene pool of the population.

For internal fertilizers, some of what determines mating success, offspring production, and eventually genetic diversity of a population happens after the male has taken off. As many of these species—marine mammals, sea turtles, sharks—also tend to be more vulnerable to exploitation, understanding all the factors that may influence successful reproduction is critical. To do this requires understanding both sides of the mating equation.

With the help of some old-fashioned dissections and some new high-tech tools, some researchers have begun to buck the trend and tease apart the myriad ways females of different species can manipulate sexual outcomes. Females exert such choice through a combination of behavior and some marvelous anatomical innovations. Vaginas, it turns out, can tell us a whole lot more than just how big a penis is.

PHYSICAL BARRIERS:
A WHALE VAGINA MONOLOGUE

It's not easy to study a whale vagina. But with a little creativity and a FedEx account, it is possible.

The vaginas of the great whales of the ocean are likely big enough to walk through. They are also potentially convoluted enough to require a GPS to navigate. That's the warning message I take from Mesnick, a marine mammal expert who specializes in reproduction and mating systems. She and colleague Dara Orbach of Texas A&M University at Galveston are taking the plunge into the little-known world of whale vaginas. Although they haven't had the chance to investigate an eight-foot-tall reproductive tract (yet!), they have looked at several species of smaller whales, dolphins, and porpoises, and what they are finding out is rather astonishing—as is the way they conduct these studies.

Studying a whale vagina is a formidable task. Just consider the logistics. Whale vagina hosts live far offshore where we can't see them, and even when we do find them it's tough to get close and challenging to stay long enough to witness them in the act. And while lots of unsuspecting snorkelers catch glimpses of dolphin penises ready for action, it is far more difficult to figure out what is happening up in their vaginas. So how do these two researchers do it?

First, they turn to the literature, in particular early naturalist studies that include prolific descriptions about the detailed anatomy of species around the world—especially these mysterious mega "fish" that occasionally washed ashore. When a specimen stranded, the local scientist would do a dissection.

"Back in those days," notes Mesnick, "scientists knew they were describing something new and different, so they took really detailed notes." Today, the researchers use these old papers like treasure maps, which lead through what turns out to be a bewildering assortment of female reproductive structures across many different species of cetaceans. Of course, sketches and notes can only get one so far; to really understand what is happening in the world of whale vaginas requires rolling up your sleeves—at least past your elbows—and diving into some dissections.

Since whales are protected under the Marine Mammal Protection Act in the United States, researchers can't just head out, catch a few, and cart them back to the lab. Instead, the team relies on a network of researchers that abide by standardized protocols for dealing with marine mammals that wash up on shore. When one strands and dies, scientists want to know why, so they conduct a necropsy (an animal autopsy). "We reach out

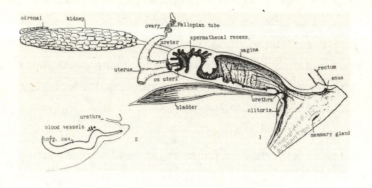

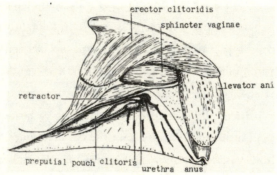

Drawings from Figures 1 and 2 in Meek, "The Reproductive Organs of Cetacea," 1918.

to people who find these animals and ask them to save the reproductive tracts and send them to us." And when Mesnick says "tracts" she means the whole kit and caboodle, from the clitoris (yup, female whales and dolphins have a clitoris!), up and around the vagina, to the uterus, the uterine horns (aka fallopian tubes), and the ovaries.

Mesnick admits that asking a stranger to FedEx a whale vagina can be awkward, even among scientists. But it works—as long as there are no mail mix-ups. Can you imagine? Some researcher down the hall excitedly starts opening a box, expecting her new microscope, and . . . what the $%*&??! She instead finds three feet of female whale reproductive parts. Nice.

Today, the researchers have a freezer full of whale reproductive tracts sent from all over. In addition to the stranding network, Mesnick also receives by-catch—animals accidently caught by fishing operations—that are collected by government fisheries observers on board commercial vessels. With a growing number of specimens in hand, Mesnick and Orbach are finding that, unlike most mammals, cetaceans have remarkably complex and convoluted vaginas. Normally, a mammalian vagina is a simple tube or cavity, with the cervix at the far end. But in some cetaceans, a series of flaps, folds, blind alleys, and funnels presents a dizzying maze for sperm to negotiate.

"It's a gauntlet. The very first dissection, when we opened up the vagina and peered in for the first time, there were so many structures in there we could not figure out how a sperm would be able to swim from one end to the other," Mesnick says. The research team has found some species have single or multiple funnels. Others have flaps or multiple folds, called "pseudocervices" because they superficially resembled the true cervix. Mesnick and Orbach are systematically documenting the variation in size, shape, and even orientation of these and other structures across species. And they've found that not every species is complicated; some have far less ornamentation. The diversity is one of the reasons these structures are so hard to study—each species looks different. Mesnick says the team has gotten to the point where they can recognize a species just by looking at the structures in the vagina. (This, of course, has me thinking about vagina-based species-identification books.)

There are several theories on why these structures exist, including that the twists and turns are designed to help keep water (which is fatal to mammalian sperm) out of the reproductive tract. But Mesnick thinks there might be more to it: "The simple question is, if all cetacean species mate in the water—which they do—and the flaps and funnels were just to keep water out, then why is there such diversity among species?"

It's a good question, and the ongoing research is helping to tease out the answer—research that is not focused solely on vaginas but includes male genitalia too. Mesnick emphasizes that it is the relationship of form and function between males and females that matters most. So she and colleagues are comparing the relative complexity of whale vaginas against the relative size of male testes in the same species. (Old publications can be helpful here too.)

Detailed male whale apparatus drawings, courtesy of Hunter and Banks, "Observations on the Structure and Oeconomy of Whales," 1787, and Meeks, "The Reproductive Organs of Cetacea," 1918.

Preliminary findings suggest that species with females hosting more complex vaginas may also be those with males sporting bigger testes, which as we've discussed, tends to indicate high levels of promiscuity. When males and females are having lots of sex with each other all the time, the females may have less control over mate selection. Instead, it is possible they could rely on more complex vaginas to do the filtering for them—during and after the act by weeding out the lesser sperm on the journey toward the egg.

At the other end of the spectrum, where only a few males monopolize the dating pool, this kind of elaborate internal female defense may not be necessary—nor such hulking testes. From the female's perspective, rather than having to select the best sperm from many males, she receives sperm from only one or perhaps a few mates. And those sperm are likely already from the most fit, pre-screened through the epic battles the males must endure in order to win access to the female in the first place. Mesnick and colleagues hypothesize that in these species with small testes, they will also find females with simpler vaginas.

These studies on both female and male genitalia indicate that whale vaginas do far more than just passively receive whale penises. They may in fact be a line of both defense and offense, preventing unwanted copulations via forked pathways and trap doors, or encouraging the fastest or otherwise most fit sperm to reach the final destination by making it easy for the little swimmers to find their way. There are several hypotheses, not all mutually exclusive, about what these structures may do. Mesnick and Orbach's research is helping to sort out the answers.

While convoluted whale vaginas are admittedly fascinating in their own right, Mesnick points out that this variation among species in their reproductive anatomy reflects important differences in reproductive strategies that likely affect how a species responds to human perturbations, such as hunting. Species in which females mate with multiple males, and sperm competition reigns, may be more resilient to disturbances than species that have highly structured social systems with more complex mating strategies.

In species where mate selection may happen via mating displays or male-male competition that sets up clear hierarchies within the population, the loss of the dominant individuals may leave females bereft of preferred mates, and thus reduce mating success. This remains speculative, but Mesnick notes that it is possible to imagine that there are advantages to female whales who mate with the strongest, oldest, or most competitively dominant males. As discussed regarding sperm whales, females may not readily engage with smaller males who cannot display or compete as impressively, or they may simply take longer to choose among mates where a hierarchy is not clear. Knowing more about how these species mate—something their vagina morphology can help us to understand—helps set realistic expectations for recovery and appropriate plans for management.

TAKING IT WITH HER: FEMALES WITH PERSONAL SPERM BANKS

It had been a long business trip. Exhausting. Finally, meetings finished, she could enjoy a few drinks at the hotel's upscale bar before heading out the next morning. That the bartender was extremely good-looking was a nice bonus. One drink turned into several, and before she knew it, the two of them were tossing back tequila shots as he closed down the place. What happened next was only natural: two attractive adults enjoyed each other's company for the night. In the morning, she woke before he did and slipped out of the room to catch her flight back home. All she left was a note: "Thanks for the fun . . ."

A few years later, after a long, dry stretch, punctuated by some truly horrific first dates, she was feeling the pressure. Work had taken over her life, years had slipped by, and now the prospect of having a child was looking slim. She decided it was time. Upon her next cycle, she let her herself become pregnant—not with a new lover, but with an old one. Alone in her apartment she closed her eyes, and, sitting quietly, gave the signal to her body to release a small amount of sperm from the internal storage tubes that sat near her uterus. Forty weeks later, she became a mother, and the bartender from that tropical night a father—though he'd likely never know it.

A PERSONAL SPERM BANK can be quite convenient and may be the reason why females from nearly all major groups of vertebrates and many invertebrates have evolved the capacity to store sperm in their reproductive tracts. From the female perspective, sperm storage means a gal can sleep with someone today and have him father offspring a few years from now—*without ever having sex with him again.* This storage capacity decouples the act of sex from the timing of fertilization, a powerful mechanism that allows females from sharks to sea turtles, sea birds, octopus, and crabs to control when they become pregnant.

There are several circumstances where such separation in time comes in handy. For species where females only ovulate at certain times of year, for example, sperm storage allows a female to mate whenever the opportunity strikes and then use that sperm when eggs ripen. Similarly, in species where males and females spend much of their time apart, personal sperm banks mean females can mate when the attractive males are around, and then fertilize eggs when conditions are best for developing young.

The technique is also particularly useful for females that lay eggs over protracted time periods, such as birds. Each egg develops separately, one after another, and there is a brief window (as little as fifteen minutes!) when fertilization can occur. Having sperm at the ready is necessary to ensure all eggs in a clutch are fertilized. This strategy may also be used by female hawksbill sea turtles, which appear to mate once at the beginning of the season and use that dose of sperm to fertilize up to five separate clutches of eggs over a ten-week laying season. In this case, sperm storage not only helps time fertilization with egg development, but may also allow the female to avoid repeated matings, which can be exhausting, and, in some cases, harmful.

As Tim Birkhead, an expert on behavioral ecology, notes in his fantastic book *Promiscuity,* sperm storage also "provides an opportunity for females to change their minds." A female that is fertilized almost immediately after copulation commits at the point of sex to who fathers her offspring; females that can store sperm can choose to dump or bypass sperm from prior one-night stands should a more attractive mate come along. In species where females may be able to keep sperm from multiple males all at once, areas of sperm storage become arenas for sperm competition. This may also increase female reproductive success as it provides a way for the most fit sperm to rise to the top. Many of the studies showing this kind of cryptic female choice and (or) sperm competition have been conducted in insects, but there are many ways sperm storage benefits female reproductive success in the sea, including helping species weather times when mates are scarce.

Such was the case for a female brownbanded bamboo shark that gave birth to a healthy young pup after nearly four years of celibacy. Cloistered away in an aquarium with two other females of her kind, the only male in the tank was a Javanese cownose ray—a distant cousin and highly improbable mate. Genetic tests revealed that the offspring was all shark, the product of a salty tryst that must have occurred at least forty-five months earlier. To date, this is the longest documented case of sperm storage followed by successful fertilization of any species of shark. Many shark species can store sperm for one or two years, but this recent discovery provides evidence that sharks may be capable of enduring even longer dry spells between mating events. This could prove a lifesaver for shark species on the decline, as stored sperm within females could potentially represent a more diverse genetic sperm bank than the local population can provide.

Genetic diversity in a population is what allows for some individuals to be naturally resistant to certain diseases, or more tolerant of warmer climates, or more adaptable to changing food supplies. The greater the genetic diversity of a population, the greater the chance that species has for making it through the wide range of both natural and human-generated threats.

Which leads me to one of the strangest side effects to consider regarding female sperm storage: it provides a way for males to sire offspring from the Great Beyond. This is much more likely to happen in species with short life spans, such as guppies, where females live over a year but males only a few months. In fact, in these small freshwater fish with internal fertilization, experiments found females with reproductive tracts filled with sperm from living males as well as sperm from males two generations back, their original creators long deceased. In this sperm competition between the living and the dead, the "ghost" sperm won out up to 25 percent of the time. Which just goes to show that "sex in the stream" can border on the paranormal. Drawing upon old sperm stores as well as new may help fish give their offspring a better chance of survival in new or changing habitats by increasing the diversity of their genetic pool.

Perhaps unfortunately, sperm storage in mammals tends to be limited to a few hours or days, likely because our high internal body temperature isn't great for sperm health. So females in some mammals have developed a different strategy. And while it doesn't rely on sperm from the dead, it does involve a supernatural-like state, known as suspended animation.

Elephant seal females employ this tactic every year. After spending a month nursing their fat-sucking pups on the beach, adult females are nearing starvation, having lost up to one-third of their body weight. It's not a great time to become pregnant. Yet, as noted earlier, this brief beach interlude is the only time when males are around. So, mate they must. And although they do technically become pregnant, the females have found a way around this taxing demand—they delay implantation of the embryo. After the newly fertilized egg divides a few times, it enters a sort of extended developmental pause—technically called diapause—and simply floats around the uterus. About three months later, after the female has had a chance to feed and fatten up herself, hormones likely trigger the embryo to fuse with the uterus, and the pregnancy progresses.

The technique works to give emaciated mamas a chance to recoup. It also means females can continue to give birth at the same time every year, in synch with other females and the males' migration patterns.

WHO'S YOUR DADDY?
POST-MATING SPERM SELECTION

There are other ways females can control a sperm's sexual destiny besides storage tanks. Sometimes, it's all about giving a little extra TLC to a favored mate. To stray back into the freshwater streams for a moment, guppies once again provide an intriguing example. In one species, a female can control how much sperm a male may deposit inside her depending on how he measures up on her relative scale of attractiveness. If he arrives just after she's seen a brighter male swimming by, he's probably going to get the short shrift if he does manage to mate; if other nearby males appear drab in comparison, however, the female may allow the comparatively brighter male to offload a sizeable deposit.

It is unknown whether the greater sperm supply happens because the female ejects sperm from less desirable males, or if they encourage greater insemination through other means, such as increasing the length of copulation. Regardless of the mechanism, the same male may find himself cut off one day and encouraged the next, depending on how he compares to recently viewed rivals.

In other species, sperm selection comes down to a bit more of a "first come, first served" policy. Though this doesn't offer as much selectivity as storing sperm might, females can still use this feature to great advantage: control which males mate first, and they can secure the genetic future of their offspring. Males that arrive late to the party miss out on the siring opportunity—which means a female doesn't have to worry about having sex with them. This kind of indifference can prove highly beneficial.

No female wants to waste her reproductive energy on offspring fathered by a suboptimal mate. Plus, copulation itself can be dangerous: females more than males bear the brunt of injuries associated with intercourse (remember those barbed penises?). If she is going to risk rough sex, it better be worth it. Female green sea turtles have at least two good tricks up their shells to deter over-eager and unwanted mates: bolt for land, which quickly encourages a male to abandon ship, or assume the "refusal" position. This is where a female hangs vertically in the water column, flippers stretched out in an iron-cross position, her plastron—or belly side—facing the male. It's the equivalent of the sharp finger waggle that some women use, often accompanied by a raised eyebrow, that says, "Don't even think about coming over here." This refusal position has been known to work even with graduate students studying mating sea turtles, who need to keep their study subjects in check while in the field. Yet deterring the advances of a single-minded horny male, especially a shark or an elephant seal, is not easy.

In general, females can expend enormous energy trying to avoid forced copulation by undesirable mates. But if a female can ensure fertilization doesn't happen, then it doesn't matter whether the suitor is suitable; she can act as a more or less cooperative mate and reduce both the burden of fighting off copulation and avoiding the cost of pregnancy.

Perhaps, just perhaps, this is the driver behind one of nature's most bizarre and frankly gruesome reproductive strategies—that of the female sand tiger shark.

A few years ago, Dr. Demian Chapman at Stony Brook University became intrigued by how the unique reproductive biology of females in this species might be driving a fierce form of sperm competition in males, and what it might mean for conservation and management. Chapman and I first met back in 1996 when we worked with Dr. Kevin Feldheim, the researcher tagging those lemon sharks in the Bahamas. Since then, Chapman has gone on to become a preeminent shark biologist who uses a combination of genetics and fieldwork to better understand shark biology and ecology, and in particular their mating systems.

Worldwide, sharks are currently undergoing severe declines primarily due to a growing demand for shark fin soup in Asia, which drives a burgeoning illegal shark fin trade. But part of the problem, as noted in "The Penis Chapter," has been a basic lack of understanding of shark mating systems.

Sharks and rays demonstrate one of the greatest ranges of reproductive strategies on the planet. Smaller species, such as cat sharks, lay eggs on the seafloor while some larger species, such as great whites, give live birth to their young. As previously discussed, all species of sharks have internal fertilization, which means females produce a relatively limited number of offspring each cycle, ranging from a handful to perhaps a few dozen pups.

Female sand tiger sharks take this to the extreme, however, giving birth to only two pups each season. If one female mates with multiple males, as is common in most species of sharks examined so far, then a female sand tiger's uterus would be like a gladiator arena, the site of some of the toughest competition you can imagine. It could be the reason why male sand tiger sharks produce what Chapman refers to as "obscene" amounts of semen: "There was so much sperm pouring out of the male sharks during dissections, my colleagues were slipping and sliding on the lab floor. They started wearing rubber boots to work."

In order to find out if escalated sperm competition was indeed driving this copious sperm production, Chapman turned to the same techniques pioneered by Feldheim—the trusted paternity test—in order to uncover how many males might be gaining access to the females. In sand tigers, during the reproductive season, each uterus typically starts off holding eight to ten embryos. But unlike any other shark (or other species as far as we know), only one embryo from each uterus survives to birth: *the one that eats all the others.*

Colorfully known as adelphophagy, which literally means "to eat one's brother," it's an extreme form of intrauterine cannibalism. In other species, developing embryos feed upon *un*fertilized eggs that the female produces throughout the gestation period. In sand tigers, it is far more grisly.

The first embryo to develop breaks out of its individual sac and actively swims around the uterus. With enormous eyes and well-developed teeth, it looks like something out of *Aliens,* a description that seems even more appropriate when its behavior is revealed. By performing dissections on pregnant females caught by beach protection nets, Chapman found that in many cases, the other embryos had deep bite marks made through their individual casings. It appears as if the dominant embryo takes a "kill now, consume later" approach to its cannibalism—it bites through the casing of its fellow siblings to kill them, and then consumes them as needed over the following weeks. In what is surely one of the most extreme instances of sibling rivalry, Chapman found one of these embryos with the tail of its brother poking out its mouth.

By taking DNA samples from the mother and all the embryos and running paternity tests, Chapman determined how many different fathers contributed to each brood, and thus the relative amount of sperm competition that might be going on. As is typical in many shark species, he found signatures for a handful of different dads. What was completely unexpected, however, was that in half of the cases, the two biggest embryos—the ones that would have actually made it into the wide world—came from the *same* father.

The number of females where this happened was far more than expected by chance. Something else is going on, and Chapman suspects it might be this:

Cannibalism by the dominant embryo in each uterus intensifies sperm competition as sperm compete to fertilize the eggs that will survive to become the hatchlings. And half the time, rather than two winners, there may be only one—a single male can sire both winning hatchlings. This helps explain why males produce colossal amounts of sperm—they want to flood the uterus so they can fertilize as many eggs as possible. Access to the earliest embryos would give males the greatest advantage as those embryos have a head start to reach hatchling size and descend upon their developing siblings.

From the female's perspective, this reproductive strategy may confer several benefits. First, it allows females to grow two pups to their enormous size, about three feet in length (adult females are only about eight feet long—as Chapman says, "It's like birthing a baby that's as tall as your thigh"). There are few predators that can take down a three-foot-long fish, so these "pups" have high survival rates. A second advantage is that such cannibalism increases the odds that only the best suitors will sire hatchlings. That is, males that can ward off other males from females, as has been witnessed in aquaria; males with sperm that can out-compete the rest; or males that father competitively superior embryos. Any or all of these "filters" work to the female's advantage. Finally, it is possible that for sand tiger sharks, the loose "first come, first served" strategy allows females to accept later advances without compromising the genetics of her offspring. She doesn't have to spend energy dodging copulations—the latecomer males simply father the fodder for her two first (and only) born.

Paternity tests continue to be powerful tools in revealing the hidden sex lives of these mysterious hunters of the deep. Studies so far show that, like the vast majority of animals on Earth, most sharks have at least some multiple paternity going in. One of the pioneers behind using paternity tests in sharks, Feldheim found that lemon sharks have extremely high rates of multiple paternity—nearly every litter having more than one father. The same appears true for the brown smoothhound shark and various other species.

Yet the genetics also show exceptions to this rule. In a separate study on a small species of hammerhead shark called bonnetheads, Chapman found the majority of litters were fathered by a single male. This was especially surprising, given that bonnetheads are known to be able to store sperm from previous matings, providing the potential for females to expand the genetic diversity of their offspring. Similar results recently have been found in other small species of sharks. Researchers don't yet know if the males have some way of blocking sperm from rivals, if females simply don't mate with more than one male, or if the female can somehow select for sperm postcopulation. Regardless of how they do it, the lack of correlation between potentially high rates of promiscuity and low rates of multiple paternity has significant implications for management.

In species such as bonnetheads and sand tiger sharks, far fewer males are contributing to the total gene pool than expected based on mating behavior and population size. This lower level of genetic diversity makes the entire population far more vulnerable.

Just to be sure you don't walk away thinking sharks—or females—are the only ones with reproductive tactics straight out of a slasher movie, there is at least one other example of extreme sperm selection—more psychological thriller than horror in nature—that is worth sharing: that of the Gulf pipefish. Close cousins, the pipefish looks like a sweet and innocent seahorse but with a straight tail. Male pipefish also have an external brood pouch on their underbelly where they nourish and protect developing embryos. In the Gulf pipefish, this pouch is transparent, so it is possible to see what's happening in this usually mysterious nether region. And what happens is this: male pipefish will sacrifice their own young in preparation for a potentially better brood down the road.

In this far-from-monogamous species, male pipefish prefer big females, especially those larger than themselves. They will engage in mating more quickly, accept more eggs from large mates, and the offspring survive better when born from exchanges involving larger females. Embryos linked to matings with small females don't do as well—most likely because their dad skimps on doling out the nutrients. Instead, he takes them for himself, bulking up at their expense in anticipation of brooding a bigger, better batch down the line. It's a cruel tactic that works to the male's—and only the male's—advantage.

Because big females are not always available, a male doesn't necessarily want to pass up on the opportunity to mate with a smaller female—something is better than nothing. But to hedge his bets, the male doesn't go all out in caring for the full brood; instead, he uses some of the female's reproductive investment to his own benefit, and to the detriment of her future lineage.

SOLICITING SEX: THREESOMES AND THE SEX APPEAL OF HIGH HEELS

Sand tiger sharks may send scientists slipping across their lab floors on copious amounts of shark semen, but the true kings of sperm production are North Atlantic right whales. With a pair of testes weighing in at a whopping one metric ton (about 2,200 pounds), they have the largest tackle on the planet—ten times bigger than a blue whale's despite having a much smaller body size. To put this in human perspective, if a 175-pound male was hung like a right whale, he would have testes more than fifty times heavier than the average male to cart around in his pants—nearly two pounds worth.

This hefty endowment represents a significant energy investment on the part of the male and hints that some serious sperm competition may be at play. With the aid of some spectacular whale watching in the Bay of Fundy, Canada, we now know this is indeed the case. And in a unique twist, the extreme level of male competition seems to be something the females encourage.

North Atlantic right whales are one of the few species of giant whales that offer a glimpse of their sex lives to us surface dwellers—and often we get far more than just a peek. During the summer months, these critically endangered whales journey to the coastal waters off Canada's eastern seaboard for a modern-day bacchanalia. The surface boils as small groups of the forty-five foot leviathans feast on the blooming plankton and frolic throughout the day. Males and females rub up against one another, bumping and jostling, diving and surfacing, churning the sea as they go.

When they aren't feeding, right whales often form SAGs—surface active groups—which frequently consist of a single female surrounded by several males. NOAA's Clapham sees the behaviors as clearly mating-related, even in midsummer, when the females supposedly can't get pregnant. These whales seem to enjoy a good time, no matter the season.

Unlike in humpback whales, the males in these SAGs are pretty mellow. Occasionally, as the groups cavort, a female will surface and roll over onto her back, exposing her broad belly to the sky. Some scientists have argued this is an "avoidance" behavior, the female's attempt to dissuade a male from trying to have sex. But Clapham thinks differently. He's seen that this position is quite the opposite of the cold shoulder; rather, it may well be an invitation for a flanking male to make good use of his flexible, and in this case, rather lengthy member.

Rolling onto his side, a male easily arcs his eight-foot-long penis in a giant bow up and into the female. As she has two sides, there are two spots from which a male can choose to engage her. Or, as Clapham witnessed on one occasion, her position allows a second male to join the fun.

"We were out tagging whales as on any other day, and someone yelled out 'Penis!' which happens sometimes. But then, someone else yelled, 'Oh my god, another one!' and that's when I turned to look. It was like a porn movie."

The female's passive stance and the reach of a right whale penis make possible the world's largest known threesome. If giant side-by-side penis bows rivaling the golden arches in size aren't impressive enough, remember that these are whales—they breathe air. So this double penetration action all happens while the participants hold their breath.

Simultaneous sperm competition explains why enormous volumes of sperm work in a male right whale's favor: nothing encourages unleashing a torrent of semen like having to fire at the same time as another male. The bigger the deluge, the greater the chance each male has of flushing out and swamping the other. As for the female, inviting such extreme competition may be one way of vetting which male has the stoutest sperm, or perhaps, the healthiest dose of semen.

Taking flight—a Mobula munkiana ray leaps out of the Gulf of California. Photograph by Octavio Aburto/iLCP

Male mandarinfish showing off. Photograph by Klaus Stiefel, Packifcklaus.com

Next page:

(top) Dressed for success—Nassau grouper exhibit several color phases when aggregating, likely to help cue spawning. Photograph by Paul Humann, courtesy REEF Grouper Moon Project

(bottom) It takes a crowd—a school of mobula rays. Photograph by Octavio Aburto/iLCP

A cuttlefish couple getting friendly. Photograph by Klaus Stiefel, pacificklaus.com

This yellowtail coris changes sex from a bright-blue spotted female (above) to a more subdued-toned male (right). Photograph ©Bryce Groark, brycegroark .com

A male octopus (upper) extends his specialized penis-arm to reach under and in to a female (below). Photograph © Bryce Groark, brycegroark.com

Nudibranchs show the art of give-and-take through simultaneous penetration. Photograph by Klaus Stiefel, pacificklaus.com

Male dolphins cruising with their tackle out. Photograph by Tim Calver ©Tim Calver, timcalver.com

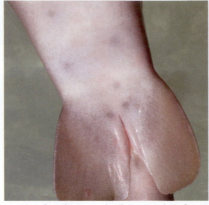

A female nurse shark pup's pelvic fins lack claspers, while a male nurse shark pup enters the world with tiny claspers. Photograph by Jillian Morris

Love bites—mating scars on a female blacktip reef shark. Photograph by Dean Grubbs

Epitoky in ragworms—bottom-dwelling Nereidid worm (left) transforms into swimming sperm or egg sac (right) to explode at surface. Photograph by Greg Rouse

Female Osedax "orange collar" lets go a string of tiny, fertilized eggs. Photograph by Greg Rouse

A male white abalone releasing sperm. Photograph by Kristin Aquilino, Bodega Marine Laboratory White Abalone Captive Breeding Program

A juvenile endangered white abalone looks out into the world thanks to the effective breeding program based out of Bodega Marine Laboratory. Photograph by Benjamin Walker, Bodega Marine Laboratory White Abalone Captive Breeding Program

A spawning burst of Nassau Grouper at an aggregation on Little Cayman, Cayman Islands.
Photograph by Jim Hellemn, courtesy REEF Grouper Moon Project

A threatened Caribbean elkhorn coral releases bundles of egg and sperm into the sea.
Photograph by Raphael Ritson-Williams

Elephant seal females also may invite sex, but under very different circumstances. After weaning her young, a female elephant seal needs to return to the sea. The distance from her spot within a harem to the water's edge may be only one hundred feet, but it's a stretch of sand patrolled by hordes of extremely unsatisfied subordinate males. Having been shunned all season, they are a living wall of lust that presents a dangerous obstacle. Especially in their exhausted postnursing condition, females are in no shape to fend off packs of up to thirty males. Even "small" males outweigh females by hundreds of pounds. As she shuffles her way toward the edge of the sea, a female has a choice: she can resist advancing males, flicking sand into their face, or sweep her hind flippers side to side in an effort to keep the male from mounting. Such resistance is risky. Males working to subdue a struggling female often bite down on the back of the neck, right where one of the main arteries lies—an artery far more exposed on a skinny female; these enormous males also slam down on the female's back in an effort to better increase the chance of penetration. Females have suffered extreme injury and death as a result of these last-ditch efforts by satellite males to sow their seeds.

Alternatively, a female may acquiesce, lying quietly as a male approaches, even inviting the leader of the pack to advance. Spreading hind flippers and arching her tailbone ever so slightly skyward, a female displays the universal "come-hither" sign in the animal kingdom: an upward raise of the rump known as lordosis. A nice pair of stilettoes imparts the same tilt of the rear, by the way, which may be the reason we find high heels so sexy.

Inevitably, the males come heaving their way over. The biggest male in the pack will throw off subordinate males, mounting the female exclusively. In return, he often will escort the female the rest of the way into the sea, perhaps mating with her several more times before she reaches waters deep enough to allow her to slip away. The female trades sex for safety, with very low risk of actually having to carry this "lesser" male's offspring—having already mated when she was most fertile with the alpha male inside the harem, the odds of fertilization by this other male are low. Whether these males have any idea they are superfluous remains hard to know.

DWARF AND DOMINATRIX:
MALES AS FEMALE SEX SLAVES

For some females, steering sperm selection isn't enough; these dominatrix dictate the fate of the whole male, not just his little swimmers.

In the black waters of the deep, a male Ceratioid anglerfish patrols the cold dark, searching for a female to sink his teeth into. Like Dracula, he relies on his keen sense of smell to home in on her distinct fragrance. For this young male, born without the ability to feed himself, finding a mate is a matter of life and death. Relying solely on the energy reserves of the egg from which he hatched, he must find a mature female before he runs out of steam. She is the key to his survival and to his posterity.

If all goes well, he will eventually make out her hulking, shadowy form as the gap between them narrows. She looms large, perhaps over ten times his size, yet he is not afraid. To the contrary, he lunges, biting her fleshy, rotund belly. But, in these waters of perpetual night, it is the vampire, not the victim, who becomes enslaved. Upon contact with her flesh, the male's mouth starts to melt as his jawbones disintegrate. Like falling into quicksand, his snout dissolves as his tissues merge with hers. Female and male blood vessels start to entwine. He can no longer see and can barely breathe, her flesh now deep down his throat. Their circulatory systems fuse, and what few internal structures he once possessed deteriorate—except one: his rapidly developing and perfectly formed testes. With his head firmly embedded into her flesh—and perhaps cued by chemicals the female releases—he begins to direct nearly all his resources into growing some sizeable sperm factories. Eventually, almost all that remains of this once-independent swimmer is a sac of sperm.

If he is the first male to latch onto the female, it is possible that his permanent enmeshing causes the female to begin producing eggs, an energy-intensive exercise she may avoid until she knows sperm are available on demand. And I mean *on demand*. As one of the early papers on this unusual mating system describes, "So perfect is the union of husband and wife that one may almost be sure that their genital glands ripen simultaneously, and it is perhaps not too fanciful to think that the female may possibly be able to control the seminal discharge of the male and to ensure that it takes place at the right time for fertilization of her eggs."

In other words, she can likely control when he ejaculates. With nothing but his rear end protruding like a wart off the side of his mistress's belly, the male anglerfish may not appear as the most macho of males; yet by becoming a permanent fixture of the female, the male has the opportunity to produce lots of offspring in an environment where the chance of finding a mate (and the energy to do so) is very small.

When those odds are even more fleeting, a more extreme form of fusion and female takeover can occur. Such is the case with *Osedax,* a deep-sea worm in which adult females entrap young larvae and turn them into sex slaves.

Back in 2003, Dr. Greg Rouse, now a professor at Scripps Institution of Oceanography, joined Bob Vrijenhoek and a team of researchers from Monterey Bay Aquarium Research Institute using a deep-sea submersible to explore the seafloor of the Monterey Bay Canyon off California. At nearly ten thousand feet below the surface, they scanned the skeleton of a dead whale and noticed many of the bones were covered in a bright red fuzz. Closer inspection and a few samples later revealed this mossy coating to be the feathery plumes of not just a new species of worm, but a whole new genus—the equivalent of finding a tiger, a lion, and a leopard when not a single species of big cat had ever been seen before.

From the Latin "os" for bone and "edax" for devour, *Osedax* defy simple definition. Today, Rouse and colleagues have identified more than twenty different species of these strange scavengers from offshore California, and more are known from other oceans, from the deepest canyons to the shallowest banks, living a lifestyle straight out of the *Twilight Zone.*

The first chapter of what I think of as The *Osedax* Chronicles begins with a description of the bizarre biology of these worms. They have no mouth or gut, so they can't eat. Instead, the adults take a play out of the plant book and grow rootlike structures with bulbous ends that protrude into the decaying bones of dead mammals on the seafloor. The roots liquefy the hard parts of the bone, and then the remaining protein or fat is passed through the skin to symbiotic bacteria inside the roots, which feast on the liquid diet. The worms then eat the bacteria for lunch. Even more strange, these roots grow from around the ovisac—like a fibrous network of food factories sprouting out of the worm equivalent of fallopian tubes.

When first deciphering the riddle of these root-bearing animals, Rouse encountered another oddity within *Osedax:* the males were missing. This was particularly puzzling as inside the females, Rouse could see thousands of tiny dots—the females were simply packed with sperm. Or so he thought.

A few months after the first expedition, while dissecting females under a high-powered microscope, Rouse noticed something strange about these sperm: they were all at different stages of development. When a male ejaculates, the sperm are all mature—ready to rock and roll. Having undeveloped sperm inside the female didn't make any sense. That's when it hit him.

The dots weren't sperm *of* the males. Those dots *were* the males.

At this point in my conversation with Rouse he paused, then added, "It was stunning, to say the least." It isn't every day one mistakes a male for a sperm. Then again, it is not every day one discovers a species where a gutless female spends her life digesting whale bones through roots sprouting from around her ovaries while amassing a harem of microscopic males 100,000 times smaller than she is. The *Osedax*'s enormous sexual dimorphism (the difference between male and female sizes) is perhaps the largest known in the animal kingdom.

When a whale dies and eventually falls to the seafloor, the current theory is that the first larval *Osedax* worms to arrive develop as females, sinking roots into the bones, growing rapidly, and producing loads of yolk-rich eggs. Later worms settling onto female-covered bones transform into males, which slip inside the females' tubes, anchor into the side, and rapidly start producing sperm. This kind of environmentally controlled sex determination, where early arriving females can dictate the fate of later arriving larvae via some kind of chemical cue, is known in other deep-sea worms and is likely the mechanism at play in *Osedax*.

The trigger to become male halts the development of all adult features, except for the testes—those bad boys are fully grown, pumping sperm through a long gonadal duct that extends from the rear and out of a pore just above the brain. *Osedax* males are basically larvae with gigantic testes that ejaculate out of their heads. This means *Osedax* joins the ranks of a few select species of anglerfish and spoon worms as part of a more exclusive club of dwarf male paedomorphs—adults that look like larvae but have mature reproductive parts.

This arrangement works well, given that males dwell inside the tubes of the females. They spend their lives living off their larval yolk sacs and pumping out sperm that can swim right down to where the eggs are produced. When the males run out of yolk, they die. So a female has to keep gathering more males throughout her lifetime. As she grows and ages, she accumulates harems of up to several hundred strong until, eventually, the food (meaning the whale bone) runs out.

Connoisseurs of decomposition, all species of *Osedax* appear to specialize in bones; all are dependent on this ephemeral and unpredictable food supply—which, as it turns out, may underlie the female's fetish for tiny males. Here's why. When food is scarce, a dwarf male lifestyle can be advantageous. Large females are needed to produce more eggs, but they also monopolize the resource. By shrinking to miniscule sperm factories that live inside the female, males can still pump out loads of sperm without having to compete for food.

Every few months new research emerges, casting light on another aspect of these complex and strange scavengers—as well as some other even stranger ones. In another type of worm, *Dinophilus gyrociliatus,* the male's fate is even more constricted. Females lay two kinds of eggs within a cocoon, large and small. The larger eggs become female, the smaller ones develop as males. The males mate with their sisters while still inside the cocoon and die soon after, never leaving the homestead. The females break out of the cocoon and live on to lay their fertilized eggs.

ULTIMATE INNER CHAMBERS

The importance of female choice and its prevalence cannot be underestimated. The more research conducted, the more evidence emerges for how females wield control over their reproductive fate.

There is one species that takes the role of female choice to a whole other level, and provides an example for just how deep (and cryptic) this force of selection might be. This incredible form of discernment is found in a group of animals that lack a true brain or most other organs of the body.

It is a comb jelly, also known as a ctenophore (pronounced *teen-o-for*), and goes by the name Beroë. Small predators, comb jellies live in plankton, with eight rows of comblike cilia that they beat in order to move. Mostly transparent, these combs are often iridescent, creating pulsating rainbows of light that ripple vertically along their sides.

In most situations, penetration of an egg by multiple sperm, in what is known as polyspermy, is fatal to the egg. It's in fact something eggs diligently work to prevent. But in Beroë this doesn't happen. Instead, several sperm wiggle their way into the egg's outer surface and into the central zone. That is when the female pronucleus—the nucleus of the egg—goes into action. Sometimes, it makes a beeline straight for one of the sperm; other times it slowly moves around the egg, checking each sperm out as it goes. It may find one it likes and immediately fuse, or it may check them all out and then return to a previous sperm as its final choice. Remember, the entire egg is just one cell—so this searching and seeking pronucleus is an organelle, a part of the cell. This would be like your liver picking out your next date.

Researchers still do not know what factors contribute to the selection process, or whether the sperm are from one or multiple males. What we do know is that female selection is operating within one of the earliest branches of the animal tree of life, at a level even more intimate than the egg-sperm interface: it's happening inside the egg itself.

Perhaps the most impressive of all inner chambers, however, doesn't have to do with those that select sperm or males—it's those that forgo them altogether. We're talking virgin birth, a process that is alive and well under the sea.

Known as facultative parthenogenesis, this is a particular type of asexual (no mate required) reproduction. It differs from other approaches to sex sans male in that it is a chosen (not obligatory) alternative to normal sexual reproduction that creates offspring genetically distinct from their mother and any siblings. This is in contrast to cloning, where individuals can bud off genetic replicas of themselves—aka clones—as a way to increase colony size or population numbers.

While such parthenogenesis has been known in reptiles and fish for decades, sharks exploded onto the scene in 2007, when Chapman and colleagues conducted tests on a bonnethead shark born in captivity to a female that had been isolated from males of her own species for over three years. But this time, there was no absentee sperm donor—instead the female had birthed her pup without any male DNA. That same year, Chapman also found evidence for parthenogenesis in a female blacktip shark, which had been in captivity for nine years, isolated from any other blacktips—male or female. When the blacktip died, she had a near-term embryo inside that proved to be of virgin birth (well, almost birthed). Since then, zebra sharks and whitespotted bamboo sharks have added their names to the list of confirmed parthenogens.

From small, egg-laying, bottom-dwelling species to large ocean-roaming requiem sharks that give live birth, the diversity of sharks that reproduce without the assistance of a mate suggests that this ability is perhaps common. But until recently, all evidence of parthenogenesis in sharks came from aquaria (and one luxury hotel in Dubai; no surprise, the same city that provides indoor ski mountains also has hotels where sharks perform virgin births). It was unknown whether female sharks used this strategy in the wild, and because DNA samples were needed from both the mother and the pups, it was nearly impossible to determine. In sharks, females aren't really the "stay at home with the kids" type. It's more a "drop 'em and ditch 'em" approach, which makes sampling mothers and babes difficult.

Until now. Feldheim, Chapman, and colleagues have developed a novel application of a genetic technique that can identify parthenogenesis out in nature. Thus far, their new tool has identified at least one species that is putting this trait to good use: the critically endangered smalltooth sawfish, which is not a fish at all but a ray.

Overfishing and loss of mangrove habitat in the twentieth century have nearly wiped out smalltooth sawfish from much of its former range in the Atlantic. Although it is a ray, sawfish fins are considered some of the best for shark fin soup and are in high demand in Asian markets; their elongate rostrum (aka nose) lined with teeth has also been a popular collector's item. Scientists estimate the population has declined by over 95 percent since the 1960s. Dramatic reductions like this set up conditions where parthenogenesis could certainly come in handy.

Since 2004 researchers have sampled, tagged, and released under strictly controlled conditions over 190 sawfish as part of an effort to better understand reproduction in this species. For example, Dr. Jim Gelsleichter, who leads the shark biology program at the University of North Florida, is heading efforts to understand the reproductive cycle of the species by studying hormone levels in blood samples collected by this program. Hormone amounts can help identify the time of year when males are producing sperm (testosterone levels are high) versus when females may be producing gametes or developing eggs (higher levels of estrogen). The ability for females to store sperm in some sharks, however, means production of sperm and eggs doesn't always coincide; determining when females become pregnant and when they give birth can be tricky.

What Gelsleichter really needs is a good sawfish pregnancy test. Sadly, despite several attempts to create one, he's come up short: "It's a reflection of just how differently elasmobranchs regulate their systems. I can't just use what works in mammals or fish. Sharks are different. That's what makes them so cool to study, but so damn hard." Instead, Gelsleichter has worked to test the use of a portable ultrasound machine that he can heft into the field and use to scan females caught during the sampling season.

Gelsleichter notes, "I look like a Ghostbuster wearing the pack. I even have goggles to reduce glare so I can see the screen."

When a female sawfish is caught, she is held alongside the boat so Gelsleichter or another research team member can lean over and run the wand over her belly, looking for pups. When a female is pregnant, it is obvious; sometimes, the researchers can even see how many pups she is carrying. Unfortunately, each machine costs upward of $16,000, so sampling is limited—which is why that pregnancy test remains the holy grail for elucidating reproductive cycles in this species.

Meanwhile, Chapman and colleagues, using DNA samples from this same program, successfully identified seven individuals, all female, all of virgin birth. Five of these were full siblings while the other two came from different mothers. This means there are at least three females out there making use of this alternative reproductive strategy, perhaps to overcome the lack of readily available mates. And it means some of the pups Gelsleichter views via ultrasound may be fatherless, further complicating efforts to identify when females and males may be mating. In some cases, they aren't!

Parthenogenesis may be a boon for smalltooth sawfish, helping them get over the hump as ongoing protections and bans on international trade take hold. The ability to reproduce without a mate could provide the extra individuals needed to keep the population humming as rates of fishing decrease. While parthenogenesis may be good in the short term, over the long haul, reliance on this strategy for survival is risky. There are consequences to skipping over males.

First, individuals that are the product of parthenogenesis have reduced genetic diversity compared to offspring born of sex between a female and a male, as they are receiving genes from only one parent. Second, offspring from parthenogenesis are almost always only one sex—whichever is the sex that has two of the same chromosome. Over generations, a skewed sex ratio in offspring and reduced genetic diversity will start to negatively impact survival of the species. Which is why, across the animal kingdom, every animal, even those that can reproduce asexually, will turn to sexual reproduction at least now and then in order to mix up the gene pool.

AT THE END OF THE DAY, sexual reproduction remains at the heart of diversity and survival of a species. And while the idea of sex by way of penis-like organ and chamber-like vagina may be familiar to us mammals, it is a relatively new invention, evolutionarily speaking.

Long before sperm squeezed through slender shafts and traveled up winding canals, the magic union of sperm and egg took place outside the body. Coming up, a look at the various ways species engage in the ancient art of external fertilization, where sex literally happens *in the sea.*

ACT II

SEALING THE DEAL

Part 2: Sexual Outercourse

UNDER THE SEA, SEX IS OFTEN AN OUT-OF-BODY EX-perience. Like trees casting pollen to the wind, the vast majority of marine species release their sperm and eggs not into or onto each other, but out into the swirling sea. Actual sex—when sperm find and fuse with eggs—is left up to these wandering gametes.

Such sexual outercourse isn't an option for us air breathers. Up on land, sperm dry out far too quickly to allow for such free-form broadcast-ing of sex cells. In the ocean, it's a different story. There, sperm can float around, bathed in all that abundant saltwater, without any risk of shrivel-ing up. That doesn't mean life is easy for these little "sea men." Finding a microscopic egg within the giant ocean is not a simple task. External ecstasy has its drawbacks, including the need to overcome the lack of a vagina.

As we recently explored, the female vagina acts as a funnel system for internal fertilizers, channeling sperm within a confined area and boosting their concentration. Sure, there are blind alleys and twists and turns dizzy-ing enough to make a poor male penis and its cargo quite disoriented, not to mention the chance of lethal secretions. Still, in terms of gathering up the troops, the female body corrals sperm to great effect. In lieu of such a containment system, external fertilizers have to rely on a different bag of tricks to triumph over the watery expanse.

The first of these is to simply pump out boatloads of sperm and eggs. Sea urchins, for example, release between ten and one hundred billion sperm with *every ejaculation.* That's two orders of magnitude more than the few hundred million sperm per spew an average human bloke can dish out. A shot glass of sea urchin sperm would hold nearly half a trillion little swimmers. Not bad for what looks like a pincushion crawling across the seafloor.

When considering sperm count, our prickly cousins are high up on the stud ladder, indeed. Marine invertebrates are so high up, in fact, that most fisheries scientists thought they were unstoppably prolific. They figured all those sperm and eggs surely met and mixed and made millions of babies. Unfortunately, this assumption overlooked the importance of density—how close together the animals are in space—in determining sexual success. This oversight has driven the collapse of countless species, from abalone to conch. Turns out, bucketloads of sperm are only one part of the strategy. In addition, ocean-bound species shrink the distance between egg and sperm by clumping up as they hump it up. Orgies, it turns out, are a fantastic strategy for sexual success in the sea.

6

OCEANIC ORGIES

Getting It On in Groups

SEX-SEA TRIVIA

- *Many orgy-loving fish like to party as the sun goes down.*
- *Oceanic orgies are so precise, they can be predicted a year in advance.*
- *Grunion add a little beach sand bondage to their orgy mix.*
- *You might owe your life to a superpower hidden in the baby blue blood of horseshoe crabs—one of the most ancient orgiers on the planet.*

SEX-SEA SOUNDTRACK

1. "Surf City"—The Beach Boys (lyrics by Jan & Dean)
2. "Let's Talk About Sex"—Salt-N-Pepa
3. "Triad"—Crosby, Stills, and Nash

True monogamy is rare. So rare that it is one of the most deviant behaviors in biology.

—Dr. Tatiana's Sex Advice to All Creation

FOR MOST PEOPLE, SEX IS A ONE-ON-ONE KIND OF thing. Not so for many denizens of the deep. From a few individuals to thousands of one's nearest neighbors, oceanic orgies are a proven practice for prolific procreation. Organizing such group activity isn't easy, though. Getting everyone to arrive at the same place, at the same time, primped and primed and ready for action—it's a logistical nightmare. Yet marine species manage to coordinate some stunning—and at times, extreme—acts of sexual synchrony.

The following daring fish, for example, push the bounds of perfectly timed erotica by adding a little asphyxiation and beach sand bondage into the mix.

FIFTY SHADES OF GRUNION RUN

Her skin shone in the moonlight, a flash of silver against the dark beach. She knew he wanted her. She could see him desperately fighting his way toward her from among the crowd. She positioned herself perfectly, knowing the site of her bound body, restrained and prostrate in the sand, would make him—would make of all them—quiver with excitement. The first to reach her, he feverishly curled himself around her. Others soon joined him, forming their own half-circle embraces. This is what she came here for—what they all had come for.

She strained with the effort, with the closeness of all those bodies squeezing against her, and yet her arousal only climbed with the number of strong, sleek bodies that writhed and wrapped around her. For miles down the beach, thousands of her sister silversides were all delighting in the same erotic embraces even as the air quickly thinned in their lungs. It was a dangerous dance, but the danger only heightened their desires.

Just when she thought she couldn't take any more, the lack of oxygen making her head spin, she felt their grips relax and the slow, sweet slide of their pleasure slipping down her body. Their desire to cling to her vanished the moment they released their burden. How quickly they relinquished their hold, turning tail and heading back into the night.

Alone, she fought to free herself from the grip of the sand, now more treacherous than tantalizing. She gasped for breath. With a final, desperate thrust, she tore free of her sandy tomb and quickly headed for the water's edge. There, she slipped back into the cool waters of her native home and let the sea pull her back to safety.

Her suitors long gone, she would never know their names, never recognize them, masked as they were by the shadow of night. But back there on the beach, tucked somewhere safely down beneath a thousand grains of sand, their DNA mingled and mixed with hers. Knowing this, she could swim at peace . . . until the moon cast its spell one more time, luring them all ashore once again.

IF FISH COULD WRITE romantic novels, the soft porn section would be filled with titillating tales of grunion runs.

In one of the most extreme acts of sex the sea provides, tens of thousands of fish hurl themselves onto the shores from the southern coast of California all the way to Baja. They are the grunion, Spanish for "grunter"—and you can hear why if you join the hundreds of visitors who flock to the shore to watch these sexual extremists flaunt their stuff year after year. Whether these folks realize they are watching one of nature's most sadomasochistic rituals—the fish cannot breathe during their shore leave, and females undertake some impressive (and risky) self-imposed bondage—is unlikely. Instead, the visitors have come to harvest a few fish by hand (the only method allowed) or simply gaze in awe at what looks like a twisting, churning river of mercury hugging the coastline.

Every spring and summer, when she is full or new, that mighty mistress of romance, the moon, pulls entire populations of these slender, cigar-length fish ashore for momentary mass beachings. Like so many other sexual acts of the ocean, this seaside orgy revolves around the moon's phases—but in the case of grunion, it's not the moonlight that matters so much as the tides.

Like any experienced surfer, grunion know that a good ride depends on perfect timing with the waves. That's why, up and down the West Coast, different populations of grunion will adjust their erotic exodus to match the precise timing of their local beach tides. They come ashore three to four nights after the highest tides following the full and new moons and repeat the affair each month during the spawning season. The window for spawning is only one to three hours—after the tide has dropped about a foot, the run will cease, and they will not come ashore again that night. They coordinate sex so precisely with the larger physical cycles of the sea that their arrival can be predicted a year in advance.

This fine-tuned environmental timing is a trait passed on to their eggs. Buried in the sand, the embryos await the right cues before hatching: likely a combination of inundation from seawater and the rocking and rolling caused by the rising tide. Thus, hatchlings emerge into the sea where they can then swim free. Though they are technically "ready" to hatch after only ten days, these newbie grunion can postpone hatching up to thirty days until the eggs are sufficiently stirred up by saltwater.

But how do the adults know when the tides are right? What cues trigger them to shuck their aquatic roots and burst forth onto terra firma? An internal clock is clearly at work, helping them gauge the season. One theory is they can sense the pressure change caused by the high tides and this, combined with changes in daily light cycles, helps them home in on the appropriate window. But we really don't know how these kinky little fish orchestrate such a perfectly executed orgy.

Though a male typically leads the charge, the female directs the show (some behaviors really are universal). Deftly arching her iridescent blue-green back, she thrusts her tail to dig a hole in the slushy sand, sinking down two to three inches until only her head sticks out. In this awkward stance she proves herself one sexy beast, and the males come a-flapping over.

As she deposits her eggs into the sandy underground, up to eight males wrap themselves around her protruding upper half. They assume this "spooning" position not to help haul her out, but to use her slippery side as a slide for their sperm. The males then bolt, leaving the female buried up to her armpits, so to speak, covered in milt. Chivalry long dead, she digs herself out of the sand and catches the next wave back to the sea. The entire event takes less than a minute, though a spawning run may last a few hours with thousands of fish flapping onshore over the course of the evening.

With the buried eggs confined by sand and sufficiently coated in sperm, thanks to the nearly direct deposit by the males, grunion successfully overcome the hurdle of getting tiny gametes to mix within a gigantic ocean. It's a successful strategy . . . as long as there is a sandy beach for them to get to.

Coastal development and rising seas are putting the squeeze on all species that turn to sandy shorelines for sex. Grunion are only one example. Female sea turtles famously heave their heft over the land to lay their eggs, and many marine mammals (seals, sea lions) haul out onto beaches for a little R&R (romping and reproduction).

Today, these animals compete with approximately half of the world's 7 billion people who now live along the coast. From fancy seaside resorts to shacks and urban slums, the space where land meets sea is becoming more and more crowded. In the United States, there are more than four hundred people per square mile living in coastal counties. Farther from shore, the average density is one-quarter of that.

And it's not just that more people are moving in—the beaches are moving out. Erosion has increased significantly, due in part to rising seas and bigger storms—effects of global climate change. Additionally, seawalls and jetties all change the natural ebb and flow of sand along the coast, altering natural beach replenishment. Unfortunately, our efforts to bulk up beaches often result in unsuitable terrain, with beach slopes that may be too steep for marine life to climb or unsuitable for nesting.

There are initiatives under way to keep the grunion grunting, but success for all species that delight in sex on the edge will depend upon global action on climate change and more comprehensive coastal management.

HORNY HORDES OF HORSESHOE CRABS

Forget the teenagers making out under the boardwalk: the real love fest at the Jersey Shore each summer happens at the water's edge. That's where hundreds of horny horseshoe crabs plow their helmeted heads into the shallow wedge between surface and sloping bottom. Though the population numbers swell along the mid-Atlantic, horseshoe crabs come to shore for sex along the entire Eastern Seaboard and Gulf of Mexico, from the Yucatan Peninsula in Mexico to Maine. Traveling several to dozens of miles, males arrive first and start patrolling the waters for incoming females with help from two of their ten eyes. The sandy shallows become a moving boulder field, which a female must squeeze her way through to the beach. There, she will dig shallow nests and lay her eggs.

But before she can reach the end zone, she will have to make it past the defensive line of hungry-for-love males, each equipped with special hook-like claws made specifically to grasp and lock onto her shell. By the time she gets to shore, she can have up to five or six males scrambling to cling onto her back. Those that cannot latch on directly appear content to hop on board the back of one of the males that does manage to get a good grip. These small chains of desperate suitors drag along behind her as she looks for a suitable spot to lay her eggs. They'll only let go after they've had a chance to fertilize a recently laid batch.

When she is finally ready (or just tired of lugging this excess baggage), a female will dig a small depression in the sand and lay a few thousand of her approximately 100,000 small green eggs. She then moves forward a few inches and lays another clutch. The males cling tight, coating the sticky eggs with sperm as she goes. Her nest and the extremely shallow water help keep her eggs from scattering too far. This, along with the near-immediate spraying of sperm by her hitchhiking boyfriends, helps to boost fertilization rates. And by having multiple males all contributing to the pot, a greater diversity of DNA reaches her eggs. Since she can't choose which males jump on her back, having several sources of sperm increases the chance that at least one of them has some suitable genes to share. Such an insurance policy may be the reason why it's worth it to haul those extra bodies up the sandy shorelines.

While there may be some benefits, toting all those males around is a burden. But there are other, greater risks associated with this mating strategy. Pick a beach where the surf is too strong, and eggs and sperm will be swept away; plow too far ashore, and females and males wind up as horseshoe crab jerky for the seagulls. And of course, there's the barrage of beaks just waiting to snatch up those freshly laid, tasty, fatty eggs. Shallow breeding likely helps reduce risk of predation from below the surface, but it only facilitates the assault from above. But with tens of thousands of eggs per female and thousands of females per beach, the strategy has worked well over the long stretch of horseshoe crab history.

Surviving mass extinctions and the shifting of the continents, horseshoe crabs have managed to find suitable beaches for their high-tide orgies since well before the dinosaurs. More closely related to scorpions and extinct trilobites than they are to crabs, horseshoe crabs look the same as their fossil ancestors did over four hundred million years ago. It's likely they reproduce just as their ancestors once did too. Talk about a family tradition.

Weathered survivors though they are, they haven't ever had to face hunters as fierce as humans. Just to be clear, most people don't eat horseshoe crabs, though their eggs are consumed by adventurous foodies willing to take on some risk with their roe: horseshoe crab eggs can contain the same neurotoxin as puffers, which has led some diners to their deathbeds. In the West, people tend to stick to using horseshoe crabs as bait for catching other species, or, in a move that would even leave Dracula squirming, we suck their blood—and use it to manufacture safer, cleaner drugs and medical devices. If you have ever received a vaccine, or hip replacement, or other internal prosthetics, and did not keel over from a deadly bacterial infection, thank horseshoe crabs. Their powder-blue blood is a marvel not just for its color, but because it is the most effective contaminant detection system we have for testing medical devices and serums. Their blood's supersensitivity to any trace of bacteria—in concentrations as low as *one part per trillion*—holds the key to our ability to determine if a formula or fake limb is clean or corrupted. Before these tests, there was no way to ensure the safety of shots and surgical implants.

The blood of the horseshoe crab is so important, there are now specialist horseshoe crab phlebotomists working for certified companies to drain blood out of about half a million living specimens each year. The operation looks like a scene from a B-grade alien abduction movie: rows of horseshoe crabs, bent into perfect pike position, attached to long narrow siphons pumping bright blue liquid into enormous jars while masked lab techs wander around with clipboards taking notes.

Like any good blood bank, these operations ideally return the donors to their environment alive—a bit woozy, perhaps, but still swimming. About 10 to 30 percent of the donors don't make it. Those that do survive take a little while to recover—as would you if about one-third of your blood had been sucked out. A recent study has found that bled females released back to the sea were less active. The scientists hypothesize that this effect could impact how frequently (if at all) these mamas returned to the shore to lay eggs and may be contributing to the population declines of recent years.

And just imagine the frustration if you were a horseshoe crab: after walking for miles on tiny legs up a constantly rising sandy slope, you finally reach the beach, find and fasten onto a female, hang on with all your might for days on end, and just before you get to sow your seeds, some lunatic vampire walking on two legs stabs you with a needle, drains your blood, and drops you back to the beach, a mere shriveled up shell of your former self. Definitely a buzz kill.

There is a legal fishery for horseshoe crabs of the eastern United States; they are used as bait primarily in the American eel and whelk fisheries. It amounts to several hundred thousand harvested horseshoe crabs a year. Unfortunately, in the past, the fishery harvested more than just the adults to offer up as breakfast: gravid females—the ones carrying eggs—are the choice pickings for eels and whelks. So a lot of horseshoe crab eggs were gobbled up as preferred bait, a preference that doesn't mix well with the idea of a sustainable horseshoe crab population.

Overharvesting of horseshoe crabs, both legal and illegal, has led to huge declines in populations all over the world. In Delaware Bay, where the largest aggregations occur, numbers are only roughly 10 to 15 percent of what they used to be. Horseshoe crabs grow slowly and take at least nine or ten years before they are able to reproduce—or recover from declines. Even if fishing were to stop, it would be decades before horseshoe crabs could make up for this 90 percent drop in population size.

Such a dramatic loss in a population affects more than just our medical needs. The millions of horseshoe crab eggs that coat the sandy shoreline also provide a much-needed food source for migrating shorebirds such as the red knot. The Eastern Seaboard is a rest stop for these mighty fliers on their more than nine-thousand-mile migration from the southern tip of South America to their breeding grounds in the Arctic. They depend on the huge volume of horseshoe crab eggs to refuel before continuing their journey. As horseshoe crab numbers plummet, so do populations of shorebirds.

Efforts to protect and boost the numbers of horseshoe crabs are under way. These include experiments with alternative bait sources and rearing horseshoe crabs in the lab as both a supply for industry and restoration purposes. Researchers have also made progress with synthetic-based bacteria sensors, which could replace the need for horseshoe crab blood. Some initiatives—such as beach protection—can help all beach-bonking species, including grunion.

As some regions limit harvest in an effort to conserve horseshoe crabs, demand (and price) increases, spurring illegal fishing activities. And there's probably nothing more easily poached than a slow-moving large crab that can't pinch and mates en masse in the shallows. Horseshoe crabs are a vulnerable lot. Although an increasing number of poachers driving ever-faster boats present a formidable challenge for resource-strapped enforcement agencies, high-profile busts in recent years help show that officials are willing to go the distance—including a helicopter chase—to nab those who illegally harvest this ancient arthropod.

Not all marine species go to such lengths, shirking the water for dry land, in order to instigate their orgy-loving affairs. Some fish prefer a romantic reef overlook instead.

GEYSERS OF GROUP GUSTO: FISH-SPAWNING AGGREGATIONS

In the tropics, the sun sinks quickly, and darkness swallows day in a single gulp. Dozens of species make use of this brief transition to cue their sexual exploits. In the southern region of Belize's barrier reef, snapper, jacks, and grouper—top predators of the reef—all make use of the same distinct promontory that juts outward, kissing the three-thousand-foot drop-off of the continental shelf. If this long winding reef was a face in profile, Gladden Spit would be the nose. At least seventeen different species flock to this striking underwater feature, where currents flow fast and furious. Each species has its own peak season for lust. Grouper—including Nassau—steam it up during the cold winter months, while snapper evidently prefer some summer lovin'. Jacks, triggerfish, and others also join the fun. All together, the multitude of fish species creates an endless cycle of orgies that rise and fall with the full moons every month of the year, all of them culminating within a few hours of sunset.

Though each species has its own variation on the theme, the sequence of group sex for snapper and grouper—the elite hunters and the most popular food fish—is similar. Their orgies tend to carry the spawners right over the edge.

In the near dusk, the smoky-gray fish form a hazy cloud, hovering only a few feet above the bottom, clustering near the lip, where reef meets open sea. As the light glows golden, fish swim in from all sides, swelling the school to thousands of fish. Their urge to bump and spawn grows as their numbers do. They become antsy. A few dart haphazardly across the crowd. Others rub up against one another.

In Nassau grouper, the entire ball starts to drift out over the rim of the reef. There they hang, suspended in the blue, as the sun begins to sink into the sea. Suddenly, from inside the pack, a dark-colored female rockets upward toward the surface. Those fish nearest to her quickly follow, and more and more join in from the sides. The swirling fish ball morphs into a cone shape, as the lead swimmer streaks higher and higher, others trailing close behind. At about twenty to thirty feet above the pack, the female, visibly swollen with eggs, lets loose a stream of her creamy-colored cargo before she falls in a gentle arc back down to the reef. The trailing males follow suit, swimming up through the cloudy residue, adding their own genes to the mix. It's an Old Faithful made of fish, with the central column shooting strongly up, up, up, and then cascading out and down. Such geysers of group sex erupt again and again from within the larger aggregation, some involving dozens of fish, others swelling to hundreds.

Brad Erisman, an assistant professor at the University of Texas at Austin, notes the chaos of a giant spawning event at Gladden Spit "gets your heart thumping, there is so much activity." First, there is the swirling mass of thousands upon thousands of fish. Then, there are the subtle yet clear courtships—changes in colors, specific darting or dashing runs. Then there are the subgroups of hundreds of fish that—after the courtship—break off and shoot skyward, releasing all those gametes into the sea.

"And it is not just a poof of sperm or eggs," adds Erisman. "Because of all the fish, it's more like a conveyor belt, one after another, the males releasing all that sperm, the females all those eggs—it looks like a volcano erupting. The fish just continue to swim upwards through this torrent of white."

The entire performance, including the final crescendo of the last fish, normally lasts less than a minute. Then, the entire group drops back down to the seafloor, regroups, and heads upcurrent to repeat the affair again. In these fish, sex is an orgy of orgies.

During the spring and summer, when the snappers spawn in full swing, thirty-foot whale sharks hover slack-jawed above the frisky fish as the strong currents wash buckets upon buckets of nutrient-rich eggs into their mouths. Like reclining Roman emperors being fed by doting servants, the whale sharks barely have to move to gorge themselves. Meanwhile, in contrast to the languid giants above, sharks and dolphins streak in from the sides to feast upon the unsuspecting revelers.

All this commotion can make it hard for a scientist to keep up, says Erisman. "The currents are strong, there are thousands of fish swirling around you, you have watch out for the whale sharks as they aren't paying any attention to you whatsoever and could just knock you out with one tail swoop. And then there are the sharks—whitetips, and bull sharks, all darting in at the fish you are trying to count. It's organized chaos."

Add to this the issue of disappearing visibility. Before the action starts, a diver can often see at least a hundred feet at the site; a few minutes after the spawning begins, visibility can drop to nearly nothing, simply from the amount of fishy goods in the water.

About thirty minutes after sunset the waters fall quiet again. The fish swim within a loose ball, once more hovering over the reef bottom. The light fades and with it the group libido. They will rest at the edge of the reef until the setting sun once again signals it's time to get the party started.

For Erisman, the real excitement of the entire ordeal lies in understanding what triggers the fish to begin their dramatic rushes toward the surface. How are they coordinating who follows whom? What kind of courtship is occurring and for how long? How often do the same fish group together—in other words, are there such things as preferred partners within a giant orgy? The buildup to successful sex depends upon the decisions and actions that precede the final rush—activities that likely change when hooks or nets are dropped into the water.

"When we insert fishing into a spawning aggregation, it disrupts all of this behavior. We know from other species, such as birds, that when we disturb these events, it can break down the system." The problem, notes Erisman, is not just the physical removal of fish. Fishing activity is likely stressing fish out, too, which can affect the hormone levels in the fish left behind. These are the hormones that control color changes and behavior associated with courtship; they help sex-changing fish gauge when to change sex, and then kick off that process.

"When you've got a bunch of fishing gear going through a spawning aggregation, it may be knocking back these courtship cues, and in effect resetting the clock, which would reduce the ability for the fish to spawn."

Around the world, the track record of fishing on spawning aggregations is pretty bleak. As documented in Nassau grouper on Little Cayman, even a little bit of fishing pressure can be too much. This heightened vulnerability is likely why about half of known Nassau grouper spawning aggregations throughout the Caribbean have been fished out, and those that remain have far fewer fish. Where there used to be tens of thousands, there are now only dozens to hundreds of fish. The biggest site hovers at around three thousand spawners. For many species that rely on annual orgies to reproduce, their formerly cascading geysers are now a mere trickle.

Although spawning aggregations are a critical part of the life cycle for the future of these fish, they are not always considered when we set fishery rules and regulations. Perhaps the most infamous example of this is the Atlantic bluefin tuna.

One of the mightiest fish in the sea, the Atlantic (also called the giant) bluefin tuna can weigh in at two thousand pounds, reach thirteen feet in length, and live for more than thirty years. They crisscross ocean basins in a period of weeks, as if swimming warm-up laps in a pool. Their sleek bodies and retractable fins provide the ultimate streamlined silhouette, while their unique circulatory system conveys the power of a warm-blooded animal to a cold-water fish. They are built for speed *and* endurance. Imagine a fish the size of a VW bus, with the speed and agility of a Ferrari, and the range of a Passat TDI. That's a bluefin. And that's why you wouldn't want to get in the way of one during a spawning rush.

From out of a swirling pack of hundreds to thousands of bullet-shaped bodies, a large female will suddenly burst free and streak toward the surface, all the power of an elite hunter channeled into a focused sprint. As she nears the finish line, where sea meets sky, she turns her massive brushed-chrome side toward the heavens and releases a stream of eggs. In a flash, she is dive-bombing the depths, rushing past a surge of males rocketing toward the surface. Following her sudden ascent, the males propel themselves upward at enormous speeds, iridescent blue lines flashing across their steely gray backs, and add their own swirling tornado of milky sperm to the churning waters at the peak of their climb. As quickly as they rise, they descend, disappearing into the dark waters below. The eruption of these one-ton missiles from the depths leaves the surface boiling.

The frothing sea and slick of sperm and eggs is a telltale sign that bluefin are below, a signal easily seen from above. For years, spotter planes, working closely with tuna boats on the water, relayed the location of spawning schools to captains who quickly encircled the thrashing giants in their nets. Such voyeurism, combined with industrialized fleets, helped fuel a billion-dollar bluefin tuna trade that over the past few decades drove their collapse.

Like a page out of a Greek tragedy, the amazing stamina and unique physique that make bluefin so fierce is also the source of their downfall. All that high-performance swimming creates a blend of muscle and fat that today makes them one of the most coveted culinary delicacies in the world.

Bluefin belly, known as *toro* or *otoro,* is revered by sushi aficionados world-wide, a modern-day fetish that fishers and fish sellers exploit. It wasn't always this way. Just a few decades ago, bluefin tuna was considered trash fish, discarded or ground up as pet food. But oh, how times have changed. Today, a thin sliver of the belly—enough for a single bite—may cost up to $25; a single fish regularly sells for hundreds of thousands in the high-end fish markets of Japan. And although the opening bids of the year are always inflated to draw publicity, in 2013, a 489-pound bluefin was sold at the Tokyo fish market for $1.76 million. One fish. Seven figures. Which is why it pays for fishers to catch every last bluefin in the sea—especially when they conveniently congregate to spawn.

For hundreds of years, Atlantic bluefin tuna have streamed through the narrow Strait of Gibraltar to gather in the warm, calm waters of the Mediterranean Sea for their annual love fest. Off the southern coast of Spain, large seine nets, called *almadrabas,* have enmeshed the migrating maters since Phoenician times. But unlike in the mid-sixteenth century, when between seventy and ninety thousand bluefin were taken in a season, in the mid-twentieth century the population had plummeted so far that these traditional fisheries caught only five to six thousand fish. A few years ago, it was even less.

Bluefin still return along the same spawning migration routes; they still group together before rising toward the surface to spawn; and we still catch them. But instead of almadrabas, there are purse seines—enormous walls of nylon mesh set out by speed boats, which can encircle an entire school of three thousand fish in one fell scoop. After surrounding the school, the fishers cinch the net closed at the bottom, like a purse. Throughout the 1990s and into the 2000s, these industrial fleets were catching nearly sixty thousand metric tons of bluefin—over twice the established quota, which was well above the scientific recommendations. Tuna populations plummeted.

As populations declined, fishers turned from the dwindling adults to the schools of juveniles. But rather than sell small fish into the market, instead, fishers now sell the captured young to "tuna ranches," which hold

the fish in pens for several months, fattening them up to kill them in optimal condition.

While the use of spotter planes is now banned in the Mediterranean
(though illegal operators still streak through the sky on occasion), fishers
can still target bluefin during their spawning romps in the Mediterranean.
Some improvements in management have been made: quotas were reduced dramatically in 2010, and held at scientifically recommended levels
through 2014. In the Gulf of Mexico, where a western sub-population of
Atlantic bluefin also spawn, targeted fishing is no longer allowed around
spawning sites. In addition, long-line gear that has typically led to high
rates of bluefin bycatch was banned during spawning season starting in
spring 2015.

While these warm-blooded fish still aren't yet in the clear, the recent
management efforts seem to have afforded Atlantic bluefin a little break.
Population numbers of the western Atlantic stock have shown a slight increase over the past decade or so; in the Mediterranean stock, small increases were first noted in the 2012 assessment. However, scientists admit
there is a fair amount of uncertainty around these estimates and thus continue to consider both populations overfished. With high demand and a
still-flourishing black market fueled by thriving illegal tuna harvests, Atlantic bluefin tuna are not yet in the clear.

The good news is that even slight upticks in bluefin and other species
demonstrate that regulations can make a difference. Add to that more and
more convincing cases that some fish are worth more alive than dead, and
the incentives for more sustainable management begin to mount. Today,
there are a growing number of alternative approaches to how we interact
with spawning aggregations that tap into our more voyeuristic, adventurous, and entrepreneurial tendencies (as opposed to our gastronomic preferences) to great effect. More on that to come.

As Erisman notes, the very characteristics that make spawning aggregations so vulnerable to exploitation—predictable timing and location, high densities of fish—also make them fantastic management tools.
Discrete in time and place, enforcement of seasonal closures on spawning

aggregations can offer a low-cost, efficient way to sustainably manage a population, and ideally boost returns for fishers who can catch more during open seasons as a population recovers.

There are other species aroused—and made vulnerable—by wild orgies under the sea. This next group, however, is extremely limited in how far they can move, if they can move at all. Which raises the question: How do you have group sex if you can't form a group? It all comes down to some impeccable timing.

7

SYNCHRONIZED SEX

A Neighborly Affair

SEX-SEA TRIVIA

- *Anyone who has successfully used IVF or is its product has sea urchins to thank.*

- *Depending on the nearness of neighbors, eggs may be prude or promiscuous.*
- *Corals have sex once a year, with their closest couple million neighbors.*
- *Good food gets sea urchins in the mood.*

SEX-SEA SOUNDTRACK

1. "That's Amore"—Dean Martin

2. "Every Sperm Is Sacred"—Monty Python

3. "Simultaneous"—Chef (South Park)

4. "Let's Do It"—Cole Porter

ADVENTURING ACROSS OCEANS, REEFS, OR EVEN just up a sloping beach in order to participate in a giant sex fest is all fine and good for those who are up for a "destination affair." But for many marine species, such epic journeys—or even small jaunts—are just not an option. For them, life is limited to a small patch of seafloor, with yearly excursions amounting to no more than the equivalent trip to the mailbox. Others, such as corals and oysters, cannot move at all. So, how to get it on when you can barely get out the door? Bring the party to you, of course. The homebodies of the sea orchestrate some outrageous orgies, but to make it work requires a whole new level of neighborly love.

THE DEVIL'S IN THE DENSITY DETAILS:
SEX BY CRITICAL MASS

It's five o'clock and the subway car begins to fill. The bright orange plastic seats disappear under gray wool suits. Everyone is quiet, faces buried behind tattered paperbacks or fixated on tiny, glowing smartphone screens. An old lady rests heavily in the seat closest to the door; a young, slender woman wedges between two mountainous men hidden behind their newspapers.

The mood is somber as the train snakes its way through dark, grimy tunnels beneath the city streets. It's not the kind of place where one would imagine an orgy erupting. But it's about to happen one stop farther on down the line. All it takes is a few more bodies. A blur of faces waiting on the platform slowly sharpens into focus as the train glides to a stop. The doors open. One. Two. Three, four, five more people squeeze aboard. Six. Seven. Somewhere around the eighth or ninth, it starts to happen: the car hits critical mass.

As the doors close, an unseen force ripples through the crowd. Coats are cast off, sweaters pulled overhead, ties untied, and skirts unzipped. With a fevered passion, these total strangers climb over, onto, and into one another in an ecstatic sexual rush that cannot be restrained. Every single individual, old and young, men and women, succumb to the overwhelming desire to join in the romp.

A few minutes later, with urges met, they slowly pick up their clothes, get dressed, and fold themselves back into their seats, slipping behind the newsprint once again. The train rolls onward, the next generation on the way.

ADMITTEDLY, SPONTANEOUS SEX on the subway is a bit of a stretch (and probably nauseating for anyone who rides the G train on a regular basis). But the idea that sex depends upon a crowd is no exaggeration—especially for the less-than-agile lot.

Take sea urchins. A few leaky males can cause an entire nearby cluster of sea urchins to unleash clouds of sperm, which they pump out through holes in the tops of their heads. As is so often the case, males tend to let loose their loads first, followed by the females. It's a familiar pattern, occurring in sea cucumbers and abalone, among others. With sea cucumbers—squishy, sausage-shaped cousins of sea stars and sea urchins—researchers think the slight delay between male and female spawning times might help increase fertilization rates. The sperm released by males forms a dense cloud just off the bottom through which the females' buoyant eggs must float on their way to the surface.

In general, bottom-dwelling, or benthic, invertebrates such as sea cucumbers, sea stars, and sea urchins don't tend to travel very far, certainly not when compared with migrating bluefin, but even compared with horseshoe crabs. (Deep-sea species may be an exception, as they have to scavenge across fair distances for sparse food supplies.) But around coastal and shallow seas, many invertebrates stick close to home—tubed feet can only take you so far. When it's time to have sex, these species huddle up with their neighbors as individuals broadcast millions (sometimes billions) of sperm and eggs into the currents.

In an ironic twist, this strategy for boosting fertilization success can also pose a significant threat: even in the enormous expanse of sea, there can be such a thing as too much sperm. For most eggs in the animal kingdom, polyspermy—multiple sperm penetrating an egg—is fatal (those choosey female Beroë ctenophores are a rare exception). To understand what's going on, we've got to dive down to the microscopic front line where sperm meets egg. The battle of the sexes rages on even at this unicellular level, and, believe it or not, sea urchins are the go-to animal for studying this kind of stuff. An extremely diverse group, sea urchins can be the size of a small brown bur or enormous, such as the softball-sized, long-spined black sea urchins, which wield four- to twelve-inch needlelike spines. They look more like a medieval weapon than the underwater lawn mowers they truly are. Impressive to behold, disastrous to touch, sea urchins can be found from the shallows to thousands of feet deep, from the warmest tropical seas to the undersea plains of the Arctic Ocean.

Google "sea urchin fertilization" and you'll find dozens of animations and videos of lone sea urchin sperm finding its way home to an egg. Why the plethora of sea urchin sex tapes? Because studying fertilization in sea urchins makes sense. They are easy to keep in aquaria, spawn on command (a quick injection of potassium chloride, aka sea urchin ejaculation juice, is all it takes), their gametes are easy to collect once in the water, and because fertilization takes place outside the body, it's much easier to observe and manipulate than, say, inside an elephant. And, not insignificantly, PETA has yet to march on behalf of captive sea urchins.

So, anyone out there who has ever received fertility treatments, any of the hundreds of thousands of couples who have successfully tried IVF, and especially any of the estimated five million or more people who now exist because of such assisted reproductive technology—thank sea urchins. All of those innovations stand on the shoulders of a basic understanding of what actually happens when sperm and egg collide. And we know that from studying sea urchins.

Our own love affair with sea urchin sex parts—the gonads are more commonly known as *uni* on sushi menus—means people have taken lots of time scrutinizing the sexual cycles of these prickly spheres. To ensure a ready and plentiful supply of the popular golden gonads, divers and sea urchin farmers must know when to "pick" the ripest ones. Ask why diners are drawn to this salty mush, and the answers probably reflect something about the taste and texture—a direct consequence of the enormous number of gametes densely packed within the gonads. *Uni* is known as an aphrodisiac, a reputation that may not be far from the truth. Sea urchin roe contains a compound remarkably similar to the euphoria-causing chemicals of marijuana, which might explain some of the allure—but most diners don't know that . . . yet. Chemists are currently hard at work creating a hybrid compound that combines the benefits of THC in marijuana with the anandamides from sea urchin eggs. The new hybrid drug could help create a longer-lasting, stronger painkiller. Ah, the scent of low tide mixed with a little skunk: I can just smell the future frat house parties now.

Regardless of the motivation, here is what we now know about sea urchin sex and the interactions between sperm and egg:

First off, good food gets sea urchins in the mood. The start of the sex season happens when the phytoplankton—tiny marine algae—bloom. This event is triggered by seasonal changes, guided by daylight and temperature, which sea urchins may also use to coordinate their spawning. By synching sex with the arrival of food, sea urchins ensure all those newly hatched larvae will have abundant snacks to graze on as they grow.

Sea urchins aren't the only ones with sex lives attuned to such "microgreens," by the way. Phytoplankton help numerous species coordinate their sexual scheduling. Oysters need an abundance of phytoplankton to build up their bulging gonads; so, as phytoplankton fill the water in spring, oysters begin to pack their shells with eggs or sperm—those transitioning from male to female stock up on both.

We have a surprisingly long history linking leafy greens and sex too. Picture for a moment the upright, vertical stalk of a romaine lettuce plant. Ahhh, you say. Then, if you have a garden, you will know that when freshly cut, the leaves ooze a milky white juice (the scientific name for lettuce, *Lactuca,* is from the Latin for "milk"). These qualities gave lettuce a sacred place in ancient Egypt, far beyond the salad plate. A distant offspring of phytoplankton, lettuce was revered as a phallic symbol and sacred food of their fertility god, Min.

Blooms of "greens" in the sea also may coincide with other favorable environmental conditions for sea urchin larvae, such as low predator abundance and water conditions that support development. Before the larvae can hatch and feast, though, they have to be conceived. To make sure that happens, sea urchins turn to an even more reliable signal to coordinate sex: their own sperm.

Depending on region, sea urchins tend to spawn for three or four months in late winter and early spring. During that time, where once scattered individuals dotted the seafloor, little groups of sea urchins start to clump together. They jostle one another, spines twitching. Then, all of a sudden, one male will start to let loose some thin wisps of sperm. As this fragrant cocktail wafts over the other males, it triggers them to all start spawning too. Sometimes, a lone sea urchin can be seen puffing away clouds of sperm into the cold sea currents. But, more often than not, spawning is a coordinated affair, triggered by the premature ejaculation of one, or perhaps a few, overeager males. The females then follow, also erupting eggs out of their domed heads like volcanoes.

What causes the first males to let loose remains unknown. But the trigger for the females is likely the amount of sperm in the water—the thicker the cloud, the greater the chance of fertilization . . . up to a point. Here's where things get tricky.

It's highly unlikely that just one sperm will arrive at the border of an egg. Like teenage boys, sperm tend to travel in packs. So while one sperm burrows down into the egg's surface, dozens of other determined wrigglers will likely try to get in there too. Similar to the opening rush in *The Hunger Games,* it's every sperm for itself in a brutal race toward the prize. But such desire is deadly for the egg. So, as soon as that first sperm fuses, the egg starts tightening its chastity belt.

When a sperm bonks into an egg, the first thing that happens is a compatibility assessment. Like an overprotective father answering the door on prom night, a layer of receptors at the egg's surface serve as a filter, gauging the merits of the suitor. Proteins at the tip of the sperm head, called bindin, either make the cut or not, depending on their match with the receptor. Sperm from a different species that bump into the egg, as can happen in the sea, are easily rejected. Even sperm from the same species may be turned away if their bindins don't measure up. It's a brutal world out there for sperm—they've got to meet the highest criteria to get a head in the door, so to speak. And the recognition proteins are the key.

Once a sperm is determined a good match, it gains entrance and can fuse with the egg. For the sperm's part, penetration is so exciting that its head explodes. Male sex parts just seem hardwired for eruption. The explosive response of the sperm helps propel the sperm inward, which then kicks the egg's metabolism into high gear—just in time.

One-tenth of a second after the first sperm dives in, an electric pulse ripples across the egg's surface, making it nearly impossible for any other incoming sperm to bind. This fast block is like a giant force field that surrounds the egg and prevents entry from the outside. Within ten seconds of sperm fusion, a second, slow block comes into play in the jellylike membrane that surrounds the delicate inner chamber of the egg. Small capsules burst forth inside the fluid membrane and turn the oozy moat into a frozen, impenetrable defense. As the membrane swells and stiffens, it first traps any embedded sperm and then squeezes them out like popping a zit. As sperm go flying off into the blue abyss (or remain locked in the stiffening membrane like insects in amber), the hardened envelope around the egg, called a fertilization membrane, continues to protect the developing embryo.

The whole process happens within a few seconds. Meanwhile, the original sperm continues to plow its way inward. It dumps its DNA, which combines with the egg's DNA, forging a new genetic blueprint. Diversity is created, along with the next generation. This sequence of events is so fundamental to life on the planet that we see it happening across the animal kingdom, from sea urchins to us, wherever sperm meets egg.

Too many sperm is only half the battle for external fertilizers, though. They've also got to deal with too few.

At the microscopic level of sperm and eggs, the ocean is an unpredictable place. In one ecstatic pulse, a female may release a group of eggs that catch a current and float off over the reef edge into a barren abyss. A moment later, her next batch may swirl over a spawning male and straight into a cloud of millions of sperm. Such treacherous conditions leave the female in a predicament: the first group of eggs die virgins, the second batch are killed by polyspermy.

What's a girl to do?

Before we explore the solution, consider this: on land, females rarely face this conundrum. Instead, when sperm are available, they tend to be available in plenty of supply. This is mostly due to the fact that sperm are energetically cheap to make compared to eggs, and males generally deposit them in bulk either up in the female, near her body, or spread thickly and directly over her eggs.

This imbalance in part drives sexual selection as Darwin first described it, with males competing for females, or on the microscopic level, sperm fighting for access to a limited store of eggs. To win these wars, as we've already discussed, males evolve weapons of seduction and competition. But for broadcast spawners—species where individuals let loose scores of egg and sperm in a giant free-for-all that requires no post-release parental care—both males and females are equally dependent on chance for the fate of their gametes. The discrepancy between numbers of eggs and sperm is not nearly as pronounced as it is for internal fertilizers. Instead, the playing field for external fertilizers is much more even. Males do not have to compete in the same way, which is why males and females look the same from the outside. Until they spawn, or are wrenched open by a prying researcher, it's impossible to tell a sea urchin lass from a lad.

Sex via broadcast spawning means choice happens at the level of sperm and egg and is dictated in part by how gregarious a population may be. In sea urchin species that live in high-density groups, females have evolved very selective eggs that restrict which sperm can get in. With many individuals living and spawning close together, there is likely an overabundance of sperm and high risk of polyspermy. Under such circumstances, it's best to be a prudish egg. But in species where individuals are more spread out, it pays to be a bit more, shall we say, open to opportunity—beggars can't be choosers. Sea urchin species at lower densities have eggs that are less discerning and more welcoming of the sperm that can find them. Species with intermediate clustering tend to have eggs that fall in the middle of this prude-to-promiscuous spectrum.

Density matters at the most basic level of sexual success. Which is why when we change the density of a population, such as through overfishing, we might be toying with some pretty fundamental aspects of sex in the sea.

Sometimes, a species can handle it. Take the red sea urchin. These common inhabitants of the west coast of North America can live to be two hundred years old. In an unusual twist of fate, we actually facilitated their population explosion over the twentieth century by wiping out their main predator: sea otters. Biology professor Don Levitan at Florida State University had the idea of using these very old but still-kicking sea urchins as time capsules to study how the sex lives of red sea urchins may have changed as their population densities increased.

Here is what he found: In the population of older sea urchins, the receptor proteins for promiscuous eggs were far more common than in the population of younger sea urchins, who had more prude proteins. It makes sense. Two hundred years ago, when the old salts were born and sea otters devoured crunchy sea urchins by the paw full, there were fewer sea urchins, and thus less sea urchin sperm around. The risk of polysperm was low. So eggs with more promiscuous proteins successfully lured in more sperm and became fertilized. Genes for these more welcoming proteins thus spread through the next generations. Contrast that with an ocean jam-packed with sea urchins, where the risk of polyspermy is high and the best proteins for eggs are prude proteins that protect eggs from too much sperm. Sea urchins with such discerning proteins likely started to have higher fertilization rates as sea otter numbers dropped off and sea urchin populations increased in density. That's why their offspring (today's young sea urchins) have higher proportions of these prude proteins than the older crowd.

The good news is that such rapid evolution of a trait (two hundred years is relatively fast for a shift that causes a change in something as fundamental as a protein) shows that a species may be able to adapt to large changes in population densities *if a winning variation in some trait is already present in the population.* In Levitan's study, a formerly rare type of sea urchin with prudish eggs likely started to have more reproductive success than its neighbors as densities increased. Thus, the gene for the prudish receptor protein increased over time in the population.

But what happens when a species declines rapidly, as has happened across the globe with sea urchins in the Mediterranean, or abalone off the west coast of North America, or conch in the Caribbean? Unfortunately, finding relic old-timers to track the changes in sperm-egg compatibility has proven difficult, leaving Levitan still searching for samples. What we do see, however, is the overwhelming failure of any of these now-rare species to rebound, even after decades of protection.

Abalone off the West Coast of the United States serve as a grim reminder that some species just can't get by without a little help from their friends. Flat-shelled marine snails, abalone have a big wide foot that sucks onto the rocky shoreline with a hold strong enough to withstand the most powerful surges along the coastline. Just like us, they come in two genders, and similar to sea urchins, their gametes may squirt out holes in the tops of their shells.

Once incredibly abundant, abalone covered the near shore intertidal of California, turning the coast into Nature's own cobblestone walkway. Beloved for their superior interior decorating style—the inside of their shell is lined with mother-of-pearl—it was that muscular foot that killed them. It tastes way better than it looks. And though that foot could hold them in place in the face of the greatest storms, they couldn't win the war against our prying hands. Opposable thumbs are a force to be reckoned with.

During the time of their decline, scientists didn't yet understand how important density was in maintaining the health of a population. It was assumed that an animal that could spew billions of sperm and eggs over multiple weeks of spawning could never be depleted. The ability of all those gametes to actually meet up and fuse wasn't ever questioned. But separate abalone by more than about three feet, and fertilization drops off precipitously. So although there were theoretically lots and lots of abalone along the West Coast pumping out millions of gametes each, they weren't close enough together for fertilization to happen—they had been thinned out by divers and couldn't make up the gap. One after another, from the pinks, to the reds, then the greens, the blacks, and finally the whites, their populations tanked. In 2001, the white abalone was added to the Endangered Species List. It was the first marine invertebrate ever to make that grade.

And they aren't alone. Other marine species suffer from what are now known as Allee effects: when population numbers drop below a certain threshold, the sexual success of each individual also tanks. This often happens because fertilization requires a certain group dynamic, a critical mass, to keep libidos humming and help wandering gametes meet and mingle. People don't suffer from this effect. The odds of finding a partner may be lower in a small town with a small, dispersed population, but once two people meet up, their chance of making a baby is the same as if they met in a big city.

To help these sensitive populations such as abalone rebound requires management strategies that focus on density, not just total number, of resident species. And that is starting to happen. For example, California closed its waters to all fishing for abalone south of San Francisco in 1997 and has a recovery and management plan in place for all species. This plan includes monitoring density, in addition to the number of new baby abalone that appear. In addition, an active captive breeding program at Bodega Marine Laboratory at University of California, Davis is rearing white abalone with the hopes of one day having enough to re-seed the diminished populations in the wild.

This focus on securing not just the right number of individuals but the right number within a specific area is especially important for the next and last group: sessile animals, individuals permanently rooted to the seafloor. These species live a lifestyle where moving to mate—even just slightly—is simply out of the question.

IMPECCABLE TIMING:
SEXUAL SUCCESS FOR THE IMMOBILE

For corals that live beneath the sea
Their days are marked by celibacy,
Excepting one night
When the moon is just right,
They engage in a million-strong orgy.
But for them to secure the prize
Corals must be ever-wise,
For there is a catch:
In order to dispatch
Their orgasms must be synchronized.

Compared to corals, we are raging sex addicts. Studies show that people between the ages of twenty-five and fifty and in a relationship are likely having sex a couple of times a month, if not weekly. In the slightly older crowd, half of men and women get it on at least a few times each year. That's a lot of sex by coral standards.

In contrast, for corals, sex is like New Year's Eve: a highly anticipated event that happens once a year and brings inflated odds of getting lucky. But, unlike the parties we're most familiar with, the chance of scoring has nothing to do with the amount of alcohol in the room. In fact, the delicate precision required for successful coral sex demands a cold sobriety. When orchestrating the largest synchronized sex act on the planet, you've got to have your wits about you.

To help wrap your head around what a coral mass spawning looks like, first imagine an apartment building full of couples all having sex. Then picture each couple climaxing at exactly the same moment, not only with each other, but *with every other couple in the building*. Now blow that image up to encompass the entire island of Manhattan.

If the cover stories of *Cosmo* are anything to go by, the majority of us may fantasize about such sexual synchrony, but few of us ever achieve such a feat—even when it's just one on one. When talking coral sex, we are talking millions of individuals, across miles of reef, all coordinating their climax within moments of one another.

Corals do all this and they don't even have a brain.

They don't really have a choice, either. Cemented to the seafloor, corals are literally stuck in the muck. Barnacles overcome such immobility with that enormous, extendable phallus that can easily play the field across the tide pool. But corals are not so generously endowed. Instead, they must rely on their gametes to do the dating and mating for them. Sea fans, sponges, bryozoans, oysters, and a host of other marine invertebrates are similarly immobile and have also developed distinct strategies to overcome their sedentary lifestyle when it comes to sex. For such an immobile lot, their best chance is to release their gametes at the same time as their neighbors, wish them well, and cross their tentacles that the night is a calm one.

Beautiful, delicate animals, corals build the only living structure that can be seen from space: the giant, winding Great Barrier Reef. Millions of tiny coral polyps collaborate to create the contoured living walls. Each polyp looks like a small fleshy blob with a tiny mouth in the middle, ringed by long, flexible tentacles, all squished inside a small, hard cup the size of a pencil eraser. There's one opening for in and out, a simple nervous system, stomach, and a full set of reproductive organs—two sets, in fact. Most corals are hermaphrodites, with each polyp producing eggs and sperm.

The reef itself is built from calcium carbonate (the source of limestone), which polyps excrete to form the tiny cups that protect their soft bodies. Depending on the species, each polyp may build its own cup, sitting side by side like a series of tract homes, or they may share walls, arranged like an apartment building. Some, such as the brain corals, are even more communal: polyps are spaced along sinewy grooves with no walls separating them within each valley.

When corals reproduce, the actual sex takes place up at the surface, where sperm and eggs collide. Similar to the concentration effect of copepod singles bars, floating eggs and sperm collect at the surface, turning a 3-D environment into a 2-D one, cramming the gametes into a thin area where they are more likely to bump into one another.

The action, however, starts several feet below, where—from amid the shadowy skyline of the reef at night, a few days after the full moon—the small gametes begin their treacherous journey toward fertilization. In corals, birth comes before conception: nonchalantly dubbed as "setting," each of the hundreds of polyps that form a coral colony starts to swell as a single, tiny sphere extrudes through the mouth. The globules look like bright pink-orange Nerds candies. Within a few seconds, what was once a relatively smooth, earth-toned dome becomes a bumpy, polka-dotted mound, as if the entire coral colony has broken out in hives.

The spheres are the bundles of egg and sperm, the seeds for the next generation of reef builders. All year long the corals store up nutrients in order to provide enough energy to produce and package these fatty balls of gametes. On the night of spawning, they slowly make their way up to the mouth of the coral, ready for release.

But, as suddenly as it all starts, the activity stops.

It is rare that stillness grips a reef, with currents in constant motion and animals of all shapes and sizes always darting to and fro. But on this night, for the briefest of moments, it's as if the reef pauses, taking a deep breath before a long exhalation. And then it happens.

All at once, across the entire reef, colonies release their bundles. A soundless pop, replicated thousands of times, and the small balls slip free. In the world of water, slow motion presides, and the release appears almost as a delayed response. The buoyant bundles, still attached to their parental polyps via thin umbilical-like strings of mucus, hover just above the coral surface.

It is in this instant that the spectacle is most magical. Swaying side to side, the hundreds of soft pink balls form a tangible aura around the coral head: all the potential for the next generation lingering for a moment, attached to the past, anticipating the future. Almost imperceptibly, the mucus strands stretch and finally break. The bundles rise in unison, floating their way to the surface like tiny balloons. Free at last.

This is the only time in their lives that corals can really move across the reef, nascent travelers on a virgin but final trek. So begins the next generation, requiring only a little mixing at the surface. *If* they get there, that is.

This is the part when the rather civilized and almost tranquil waltz of the gametes turns into a full-blown rave. With many different species of coral spawning on the same night, watching a mass coral spawn is like witnessing a snowstorm underwater, except the snowflakes fall upward, are colored bright pink, and are the energy equivalent of a Snickers bar. It's the ultimate all-you-can-eat buffet and nary a reef resident is likely to miss it. Within moments of the bundles' release, kamikaze crustaceans dive-bomb the water column in a mad dash for the fat-packed pink parcels. Tiny shrimp and worms crawl almost frantically across the corals, engulfing what bundles they can before the tasty morsels rise out of reach. As small fry gorge themselves, bigger fish swoop in on the preoccupied diners. It is a night of indulgence for all, and Nature ensures nothing goes to waste, right down to the fish poop—the excess from the night's feast—which is gobbled up by microbes on the reef. But even the hungriest fish becomes full eventually. And with all those tasty gametes and fatty mucus floating around, enough lucky sperm and eggs can make it through unscathed.

Spewing forth gametes on the same night as other species helps to swamp predators, but it also brings some serious risk: parents have little control over which of their gametes will meet and mingle with their own species versus a totally different one. Thanks to internal fertilization, we can be pretty sure of who, and especially which species, our gametes are associating with. For corals, there are dozens of other closely related species clustered nearby, often going off on the same night. Their gametes swirl around in a sea of truly mixed company. And that means they are at risk of forming hybrids.

What's so bad about hybrids? Well, it's okay if you are a mule. Mules are the product of a female horse crossed with a male donkey—as opposed to a hinny, which is the offspring of a male horse and a female donkey (it's amazing the trivia one learns when researching animal sex). But, like mules, most hybrids are sterile or unviable. They are genetic dead-ends. Making a hybrid kind of defeats the main purpose of having sex in the first place—that is, to pass one's own genes on to future generations.

While some hybrids can reproduce and even thrive, their offspring, within one or two generations, tend to falter. In the world of hybrids, you are either a stud, a dud, or, worst of all, a stud succeeded by duds. The last may lead to a particularly vigorous hybrid outcompeting its parents, only to die out a few generations down the line. The result is the loss of all three species—the two parents and the hybrid.

So hybrids are something most species try to avoid. Corals reduce the risk of hybrids by coordinating their spawning so precisely with fellow species members that by the time other species spawn, their own gametes have most likely already met and fused into healthy, fertile coral larvae.

Take the two sister (very closely related) species, lobed star coral (*Orbicella annularis*) and boulder star coral (*O. franksi,* called "franks" by those who know them well). These species are so closely related that scientists realized they were separate species only a few years ago. But you can tell them apart in the field. Lobed star coral form thick columns with bulging tops, like clusters of two-foot-tall pale-pink or yellow mushrooms. They tend to live in slightly shallower water than the franks, which look like lumpy, dark-brown mounds of Rocky Road ice cream. In addition to their looks, these two species have another distinguishing quality: they spawn at slightly different times on the same night. The franks are the early birds, going off about two hours after sunset. Lobed star coral then follow, approximately one and a half hours later.

These staggered and extremely accurate spawn times are not only impressive—they are necessary for the survival of each distinct species. The synchronized spawn helps sperm find eggs; the time lag helps ensure the right sperm find the right eggs. Without the time lag, the two species would more easily fuse.

We know this because we have tricked them into spawning at the same time and seen it happen. I say "we" because, admittedly, in the name of science, I have screwed with coral spawning. Manipulating coral sex is surprisingly easy to do. All you need is a bucket and some big black garbage bags. But there are no asphyxiation fetishes here—leave that to the grunion. Instead, the garbage bags simply work as a stand-in for the cloak of night. Toss a bag over a bucket holding some corals an hour before sunset, and lo and behold, the corals spawn exactly an hour earlier than they normally would. Do this with lobed star corals and you can make them spawn at the same time as the franks.

Coral sexperts such as Nicole Fogarty, of Nova Southeastern University, use the trick in order to test how different coral species mesh. Fogarty has spent many a night playing coral matchmaker. To reproduce nature's most intimate act she uses a long pipette—a glorified eye dropper that can suck up different-sized particles with great precision—and a whole bunch of urine cups. Fogarty uses the cups to make little sperm and egg cocktails, slurping up sperm from one bucket and mixing them with eggs from another bucket, in all possible combinations.

Through these kinds of "fertilization trials," scientists have discovered that under the right circumstances, lobed star corals and franks are compatible: the sperm and eggs of each can fertilize the other. But, as any online dating site can attest, compatibility is no guarantee of a successful hookup. Although lobed star coral sperm can fertilize a frank egg, it only happens if the frank eggs have no other choice—an "Only if you were the last sperm in the ocean" kind of thing.

Part of the reason for this isn't just the high morals of the frank eggs— it's the gusto of the frank sperm. In most species of corals, the chances of fertilization are only good for about an hour. After that, sperm just seem to tucker out. But frank sperm act like a good racehorse: alone on the track, they don't do much; add in some competition and the suckers can run.

So if franks and lobed star corals spawned at the same time, frank sperm would kick into high gear and be able to not only fertilize the major- ity of frank eggs but also fertilize the lobed star eggs, which seem far less discerning about whom they let in. And if some frank eggs were to drift into thicker, denser clouds of lobed star sperm, they too could be fertilized.

Timing, along with some innate preferences, helps keep these two spe- cies from intermixing.

So how do corals get this timing so right? The garbage bag experi- ments helped to show that the time of sunset is one big factor. The work- ing theory is this: either solar irradiance or wind fields cue the month for spawning, then the lunar cycle dictates the day. The sunset triggers the polyps to extrude the bundles toward the opening of their mouth. Just how corals sense these subtle changes in weather and light patterns—remem- ber, they don't have eyes or brains—remains a mystery.

Similar to corals, oyster sex requires alignment of multiple factors too. As noted, seasonal changes that bring an abundance of phytoplankton pro- vide the nourishment oysters require to bulk up their gonads. But tempera- ture is also critical—as water temperature rises, it warms up the oysters so they can take advantage of those nutrients and kick into high-gear gamete production. As their gonads ripen, oysters then release chemical cues, as if checking in to see how everyone else around them is doing. For oysters, the chemical cues are a kind of "call and response" across the reef, says Juliana Harding at Coastal Carolina University; she specializes in oyster reproduction. Oysters that receive the signal and are also ripe will release their own chemical cues in response: "Roger that. We've got gametes locked and loaded here too."

Tides are also important, as well as the shape and structure of the oyster reef itself, both of which influence how water flows over the bottom: since oysters can't move, they rely on the swirling currents to carry their messages—and their gametes—from one oyster to the next. Finally, oysters prefer to have sex at certain times of day, namely dusk or dawn, when it's harder for egg-munching predators to see their prey. When all of these things align, Harding says, the oysters will start.

"When spawning does begin, I have seen it proceed sort of like a human 'wave' moving around a stadium. One group of oysters releases gametes, and then the oysters immediately downstream release, and so on."

In corals, genetic factors and density—the influence of neighboring corals—most likely fine-tune the actual time of bundle release. Corals have the nifty ability to clone themselves and use this technique to bud new, genetically identical polyps that stay attached to the parent colony—that's how a colony grows in size. Every now and then a fragment may break off, forming a new, genetically identical but physically separate colony, similar to a cutting from a plant growing into a whole new tree.

Genetically identical clones spawn more closely together in time than nonclone mates. But this only works if the clones are near one another. At distances of more than fifteen feet, proximity trumps paternity and neighbors spawn more in synch, regardless of relationship. Though no chemical compound has been identified yet, it is likely that pheromones of some kind are released by spawning polyps that then cue nearby colonies to join in, similar to oysters. Such chemical cues work in other species, too, including sea urchins and abalone.

Sunset. Moon phase. Neighborly love. It all sounds rather Woodstock. The truth, of course, is that coral spawning is a highly regimented affair. In contrast to oysters, which continuously spawn throughout an extended season, a coral has just one night. They cannot afford to cast precious seed into a barren blue abyss nor risk fertilization by the wrong species' spew. Impeccable timing is truly a must.

So is having enough corals. And that is where we have started to derail Nature's coral calculus. Because corals cannot move, they cannot compensate for population declines by grouping together the way that mobile species can. When coral colonies thin out on a reef, fertilization success declines not just because there are fewer individuals spawning, but also because the increased gap between colonies likely throws off the timing of the spawn. Just fifteen minutes too late or early can cause significantly less fertilization success.

When corals die on a reef, it's thus a double whammy. And corals have been dying a lot lately.

Globally, coral reefs are one of the most endangered ecosystems on the planet, with a staggering 10 percent of all reefs already permanently lost and another 30 percent expected to decline in the next few decades. In 2012, sixty-six species were proposed for the Endangered or Threatened Species List, including the star corals. The Great Barrier Reef has lost over half of its coral cover in the last twenty-seven years; Caribbean reefs have an average of 15 to 20 percent coral cover left.

The transition from reef to rubble is an exercise in extreme contrast: an ancient civilization, great temples swallowed by jungle; a city skyline leveled to dust. And like the collapse of any thriving metropolis, the loss of coral reefs hits people—yes, us—where it hurts the most: our pocketbooks and stomachs. Coral reefs provide employment and the primary source of protein for hundreds of millions of people worldwide. On a global scale, through tourism, construction materials, fishing, pharmaceuticals, and coastal protection services, coral reefs provide an estimated $375 billion annually. That's more than the GDP of Denmark or Thailand.

None of these services or products could exist without corals. The tropical oceans where reefs thrive are, in reality, underwater deserts. That's why the water is so clear: there's nothing in it. Corals survive because of a remarkable partnership with the microscopic algae called zooxanthellae (commonly called "zoox," rhymes with "spokes"). They live inside the coral tissue and are mini–food factories, converting sunlight into energy. They pass that energy on to corals that, in return, feed the zoox leftover waste products critical to their survival. It's an extremely tight recycling system that gives corals the extra boost they need to build their thick skeletons and manufacture energy-rich gametes.

This is how corals can turn a tropical sea desert into an oasis of life. Covering less than 1 percent of the ocean, coral reefs are home to an estimated one-quarter of all marine species. They contain more species in a given area than any other marine habitat. All this life draws the tourists, fishers, and pharmaceutical giants—the last of which have only just begun to explore the bountiful biochemical resources produced by reef organisms, including many cancer-fighting compounds.

All of these benefits and resources depend on corals building healthy reefs. To do that, corals need to successfully reproduce.

And cloning doesn't cut it. Although coral colonies can grow and at times create new colonies via asexual reproduction, cloned colonies are all equally susceptible to disease, predation, and competitors. A reef full of clones is a recipe for disaster in terms of long-term survival. Instead, a healthy, long-lasting reef requires many genetically diverse colonies. That way, no matter the disturbance, the odds are a few colonies will have what it takes to make it through. And what's the most powerful way to create genetic diversity? You guessed it: sex.

For coral sex to be successful requires two things: lots of energy and seriously synchronized spawning. In the tropical ocean desert, finding enough energy is hard when conditions are optimal, and conditions have not been optimal for a long time. Sedimentation, pollution, overfishing, and climate change all create stressful environments in which corals must struggle to survive.

Sediment (often carrying herbicides, pesticides, or oils) washes into coastal seas due to deforestation and generally poor land management. As we tear down mangroves and salt marshes and build paved roads and houses, more dirt and grit washes into coastal waters. To remove sediment, corals, like us, make snot—the universal compound for cleansing the body—to trap sediment grains and then slough it off their surfaces. But making mucus pulls energy away from making sperm and eggs. Exposure to heavy metals, herbicides, and other pollutants can reduce the number of gametes and cause infertility or decreased egg size, sometimes for years after exposure. These impacts, especially if they happen in combination, make coral sex less likely and less productive. Similar impacts are snuffing out the sexual success of oysters too.

For any marine species, the more the struggle, the fewer the resources left for reproduction. The same thing happens to us: couples under extreme stress often have trouble conceiving—or for the woman, carrying a pregnancy to term. Stress and reproduction just don't mix.

Happily, corals and oysters haven't yet given up the ghost. In fact, very few extinctions have occurred in the sea, compared with on land. There is still lots of potential to work with. Teams of scientists and snorkelers, reef managers and entrepreneurs all continue to search for ways to reverse declines in corals and oyster reefs. Fields of farmed coral, mounds of recycled oyster shells, the upward trend in large, fully protected marine areas, protection of herbivores, and other efforts are all helping turn the tide. And because corals and oysters are ecosystem engineers, solutions that assist them can also help boost the sex lives of other species too.

POST-CLIMAX

SEX HAS BEEN THE FORCE FOR PROPAGATION IN THE sea for upward of a billion years, at least. And it will continue to be so. But *what* it propagates, and *how much* it propagates, is, in part, up to us to decide. We affect the fate of the sea, just as the fate of the sea affects us. So far, our actions have tended to spur proliferation of the more gelatinous, slimy, and microbial in nature. But it doesn't have to be this way.

Generating recoveries for depleted species and ecosystems and preventing future declines ultimately depends on understanding how our ancient salty cousins of the sea reproduce. The more we learn, the better the chance we have to adjust our own behaviors and find accommodation between marine life and ourselves. As the final chapter illustrates, we are already well on our way.

— 8 —

TURNING UP THE
SEX DRIVE

*How to Spark Successful
Sex in the Sea*

SEX-SEA SOUNDTRACK

1. "I Can See Clearly Now"—Johnny Nash
2. "As Time Goes By"—Herman Hupfeld
3. "(Your Love Keeps Lifting Me) Higher and Higher"—Jackie Wilson

RIGHT NOW, SOMEWHERE OUT THERE IN THAT BIG blue sea, you can bet that some creative—if not sneaky, and perhaps rather acrobatic—hanky-panky is going on. Though we may not always be able to see it, we've lifted the veil enough to know a colorful and kinky world of sex drives the ever-unfolding diversity and abundance of life in the sea.

As any good sex ed course will tell you, sexual behavior has risks and rewards. In the case of sex in the sea, however, the risks we need to worry about are *to* sex, rather than *from* sex. A lot can go wrong during the buildup to and performance of this final act. We have all been there. The soft, romantic light, the perfect music playing, you and your crush, alone. You lean in for that first kiss and just before those proverbial fireworks explode . . . BAM! your roommate bursts into the room flicking on all the lights.

It is no different in the sea, where a surprise attack by a predator can put a serious damper on a snapper spawning rush, or a massive storm can create too much surge for two seahorses to parade and pirouette. Such interruptions are expected, and species have evolved to deal with the occasional mood killer. But human activities have become more than an every-now-and-then disturbance to sex in the sea.

We are a major prophylactic.

That is a problem, and not just for frustrated fish. We are far more connected with the intimate acts of our water-bound cousins than we may realize, for better or for worse. For better, because all that sex in the sea provides valuable products and services that boost our own well-being. For worse, because this intimate relationship is a two-way affair, and our actions are currently bungling the critical dating and mating games in the deep. Despite all that potential procreation, the oceans are far emptier today of our favorite seafood and colorful coral reefs and far more full of plastic particles and dead zones—areas with too little oxygen in the water for fish to survive. From a human-centric perspective, the global decline of ocean life has a negative impact on local economies, food security, shore-line stability, and water quality. It deprives us of medicinal resources, as well as recreational, spiritual, and cultural values. This is what is at risk should sex in the sea continue to slump.

But don't be too deflated. The good news is that overfishing, climate change, pollution, and other human-caused mood killers have yet to per-manently put out the sexual fires of the sea. As dire as some of the statistics may be, this is, in fact, a time of great optimism for the future of the ocean (there's even a hashtag for it: #oceanoptimism). We sit poised at the cusp of a potentially new relationship, one in which we can choose to support the sex drive of the sea, and in so doing, fuel the bounty that supports us all.

The following sections detail some of the major ocean wins in policy, new technologies, and shifts in behavior that are turning the tide of crash-ing libidos and stymied sexual affairs, especially for those species we value and depend upon most. From global trends to local actions, much is being done to create conditions more conducive to successful sex in the sea.

ALLEVIATE PERFORMANCE ANXIETY:
TAKING PRESSURE OFF THE OVERFISHED

It is tough to perform under pressure. For many marine species our expec-tations of reproductive performance have been simply unrealistic. In the last 150 years, we've increased the power and size of fishing fleets as well as the efficiency of fishermen to the point that species from snapper to sharks can't keep their numbers up. Their strategies for sex render them unable to reproduce fast enough to satiate our appetite for more and more seafood. Supply is dropping and demand is going up. In addition, a historic focus on catching the biggest fish has tended to leave females with scrawnier suit-ors and populations with fewer BOFFFFs, further stifling successful sex.

And don't forget about the potential for shrinkage—another downside to size-selective harvesting of the biggest fish.

Both biology and behavior play important roles in determining how much pressure a species can take before it simply cannot replace the number of individuals removed from the population. The crux of sustainability lies in finding this balance. In terms of *biology,* slow-growing, long-lived species—such as sharks and rays, sea turtles, marine mammals, as well as some fish—cannot withstand the same amount of fishing pressure as, say, sardines or some types of squid. In the case of sharks, for example, females of most large species carry at most about a dozen pups at a time, and often reproduce only every other year or every three years. While not all shark fisheries are as unsustainable, the pressure on large sharks has resulted in worldwide population declines.

At the same time, certain sex-driven *behaviors* such as forming spawning aggregations or schooling around seamounts or having particularly discerning females also make some species more vulnerable than others.

So, what can we do to alleviate the pressure?

From the supply side, one thing fisheries managers can do is pass regulations that focus fishing pressure on the middle-of-the-road-sized fish. Protecting both the babies and the big adults of a population helps ensure new adults can join the breeding pool and reduces the chance that fishing will cause detrimental change in growth rates or other important survival traits. In experiments where selective fishing for the biggest fish was stopped, populations did increase, with subsequent generations regaining some of their pre-fished girth. The recovery may be slow, but it is possible.

To enhance the opportunity for ripe and ready adult fish to get together, fisheries managers can also protect spawning aggregations with the same rigor as we do nesting colonies of sea birds. Because spawning aggregations concentrate fish in such abnormally high densities, it is especially hard to "see" declines as they are happening—aggregations look fine right up to the point of collapse. The importance of these yearly events to the future growth of a population, and the dismal track record of fishing on aggregations in the past, is enough to warrant proactive bans on all oceanic orgies of this nature. Enforcing this strategy, while not easy, already has proven effective for both fish and fishers.

The case of red hinds, a small species of grouper, in St. Thomas, US Virgin Islands, is one example. A properly enforced permanent closure of the spawning aggregation in 1999 has led to strong signs of recovery—a balanced sex ratio and double the density of fish in the local population. In addition, fish in the spawning site and outside the protected area are larger than they were before the ban, offering commercial fishers increased value of their catch.

Protection of spawning aggregations doesn't always have to come from the top down, either. In Fiji, for example, a grassroots campaign called 4FJ is under way to encourage people to give up eating vulnerable grouper species during their spawning season. According to the website, over four thousand people have pledged to not eat, buy, or sell grouper from June to September. Those who pledge are sharing stories about what a future full of fish means to them, and can work with 4FJ to bring others onboard the effort. This initiative continues to gain momentum, helping to reverse the devastating losses that have left fishers with a 70 percent decline in catch over the last thirty years.

Meanwhile, here in the United States, we can similarly help take pressure off overtaxed populations by avoiding vulnerable species and becoming a bit more adventurous with our own seafood selections. Especially in America, we tend to turn to the same species—such as tuna, shrimp, and salmon—all the time. Our limited preferences put untold pressure on these wild species to perform and tend to support the least environmentally friendly aquaculture. Meanwhile, we miss some excellent opportunities for an extremely satisfying, and perhaps more exciting, culinary experience. The ability to diversify our palate has never been easier, as chefs increasingly embrace the opportunity to be both more creative and more sustainable with their menus and new market channels provide multiple ways to buy more kinds of seafood.

One of the visionaries behind this effort is Barton Seaver, a chef and National Geographic fellow who has proven that using lesser-known varieties of seafood in creative ways can work—in terms of taste, sustainability, and economics. One night a few years ago, Seaver's local supplier unexpectedly delivered a box of flying fish—the day's bait—rather than the day's anticipated catch to his popular DC restaurant, Hook. When pressed for an explanation, the supplier simply said it was a "bad day" for fishing. So, Seaver whipped up a delicious sauce, instructed his wait staff to tell the story about the fish, and promptly sold out of the dish by seven p.m. that night—at $26 per plate.

For Seaver, when choosing seafood it's more important that we ask what the oceans can sustainably provide, rather than demanding of them only what we want to eat. It's a paradigm shift that could serve to redistribute fishing efforts to more responsible (and sex-friendly) harvests—spreading the demand of millions of hungry human mouths across more diverse and robust types of marine protein. This includes underutilized species that are often tossed away as bycatch (such as plentiful redfish or pollock in New England), invasive species that need culling (lionfish, anyone?), and what some consider "ugly fish" that may not look that appetizing but taste good (geoducks, a clam that looks like a giant phallus, apparently are all the rage in Asia).

Happily, when we do cut species some slack and provide space for them to recover, populations often will bounce back. In the United States, two-thirds of once-overfished populations have been successfully rebuilt or are nearly rebuilt since the inception of strong federal management plans in 1996. Once near the brink, striped bass off the East Coast and giant sea bass off the West Coast rebounded under strict local management and enforcement regimes.

Diversifying which species we demand helps give depleted stocks more chance to recover while still supporting fishers who can make a living targeting other species. As these species rebuild, fishers' incomes can increase and menu selections can broaden even further.

But how do we ensure that we are replacing overfished, vulnerable species with those that can take a bit of well-managed fishing pressure? In other words, how can we support those fisheries that fish in a safe-for-sex manner? It starts with getting to know our fish a bit better.

GET THE STORY: SEX-FRIENDLY
SEAFOOD "COMES WITH A TALE"

Part of the calculation for which species may be best to try out depends on how often, how abundantly, and how successfully they reproduce—in addition to how well they are managed. Sustainable seafood certification programs such as Marine Stewardship Council (MSC) and seafood guide programs such as Monterey Bay Aquarium's Seafood Watch program, include consideration of the sex lives of species as part of their assessment and are a good place to start. For example, does the species take a long time to reach sexual maturity or reproduce early and often? Do fishers target spawning aggregations or abide by seasonal closures? As a general rule, species lower down on the food chain (think small fish such as sardines and shellfish) tend to reproduce more quickly and abundantly and are also good choices.

The biggest challenge to supporting responsible fishing, though, is that the vast majority of fish that make it to our plates are, in reality, a complete mystery—a filet served up without any tale. Today, it is very easy and far too common for illegal, mislabeled, and unsustainable seafood to slip into the supply chain and onto our plates. This includes fish and invertebrates harvested illegally during their spawning season, for example.

If you were to sit down at a restaurant in North America with two friends and all order seafood dishes, one of you would likely be served a mislabeled fish. This means the high-priced "catch of the day" could very well be a cheap tilapia. On an even more disturbing level, it also means the farmed shrimp in your dragon roll may have been fed fish caught by slaves at sea. Human trafficking and other social ills pervade the global seafood industry to an extent that is only now starting to be exposed. These widespread criminal behaviors have even been found in the supply chains of major grocery stores such as Walmart, Costco, and Tesco.

Between $10 billion and $23 billion worth of illegal, unreported, and unregulated (IUU) fish make their way through the global seafood supply chain every year, representing up to 40 percent of the total catch in some fisheries. Whether it is illegal fish or mislabeled fish, at the heart of the problem is a food production system that turns what should be a premium protein into a mystery commodity. With so much hidden, it makes it difficult to know if the seafood on your plate caused some serious coitus interruptus in the deep or not.

One way to ensure your seafood is not only sex-friendly but also socially and environmentally responsible is to demand what the nonprofit Future of Fish calls "Storied Fish"—fish that comes with the information about its journey from water to plate (how, where, and when the species was caught.) It's fish that "comes with a tale." Today, this is a whole lot easier to do than it was a few years ago, thanks to several new initiatives that provide more detailed information about the source of your seafood. Take Community Supported Fisheries (CSFs), for one. Modeled after the more well-known Community Supported Agriculture (CSAs) efforts, CSFs are popping up in coastal towns around North America, providing consumers an opportunity to support local, sustainable fishing efforts, including those that abide by closed seasons, size limits, and quotas. By purchasing directly from fishermen, consumers can learn the story behind their fish. CSFs have also been shown to reduce the overall carbon footprint of seafood purchases and provide a more diverse selection than that available in local stores.

Cutting-edge technology makes storied fish possible by providing new ways for sustainable seafood to be recognized and verified at the seafood counter or in a restaurant. In San Diego, for example, chef Rob Ruiz created edible bar codes for his sushi rolls. They are made from rice paper and soy-based ink, and diners can scan the codes with their smartphones and read up on the backstory behind the tuna in their spicy tuna rolls prior to dipping them in soy sauce. For those who like seafood at home, several companies, such as ThisFish in Canada and Followfish in the European Union, offer seafood selections that come with similar codes printed on the package—scan and discover the name of not only your seafood, but the boat that caught your meal-to-be. The Gulf Wild brand provides the same kind of information for over a dozen species from the Gulf of Mexico. With gill tags on the fish linked to information about the catch, the system ensures each fish is from fishers in the program that have committed to responsible harvesting practices.

PEEK BEHIND THE CURTAIN:
SEX TECH FOR TRANSPARENCY
AND TRACEABILITY

And that's not the only fancy technology helping to drive more sex in the sea. Don't worry, we aren't delving into any discussions about seaworthy sex toys; this section highlights efforts to ensure the seafood you eat is what you think it is and to prevent illegal fish from landing on your plate.

For storied fish to be possible, reliable systems must be in place to track and trace the story of the seafood through the supply chain, from where it was raised or caught to its end destination on your plate. This is the behind-the-scenes work that nonprofits and industry groups are currently tackling in order to root out fraudulent mislabeling, illegal catch, and slavery from seafood supply chains. It also allows us, as consumers, to distinguish and reward—by paying premium prices or just loyally consuming—responsibly caught fish from responsible fishers.

Historically, traceability has been one of the greatest hurdles to more sustainable management of fisheries: a notoriously old-fashioned seafood supply chain still relies on transactions recorded on paper and fails to pass information along the long, winding trail of globally traded seafood. New traceability technology systems are changing this landscape by providing verifiable, transparent records of the personal history, so to speak, of seafood. It's like a basic background check: these systems ensure that product marketed as "sustainable" isn't hiding any nasty secrets. While there is much yet to be done, the fact is that right now industry is engaged, technology companies are engaged, and progress is being made. This has never happened before, giving fish-heads like me a reason for real ocean optimism.

REIMAGINE AQUACULTURE: SEX-FRIENDLY SEAFOOD CAN ALSO COME FROM A FARM

New developments in sustainable aquaculture can also help ease the burden on wild species. A former fisher and the founder of Thimble Island Oysters, Bren Smith has started what he calls 3-D ocean farming. By suspending from ropes at the surface fast-growing seaweeds such as kelp, interspersed with shellfish such as oysters, mussels, and clams, his farm cleans the local waters while helping shellfish avoid the sedimentation that can bury them on the bottom. The kelp absorbs carbon dioxide, becoming a carbon storage tank while also filtering nitrogen and other pollutants out of the water. The ropes and bags of suspended shellfish also change the water flow around the farm, helping to create a safe habitat for juvenile fish.

The result is a farm that restores the environment as it produces a whole line of seafood products. Working with local chefs, Smith has developed innovative oyster dishes, kelp noodles, and even kelp ice cream and cocktails. In collaboration with researchers at Yale and the University of Connecticut, his farm serves as a living laboratory for testing new ocean farming methods to turn food production systems into habitat restoration systems as well. Finally, through the nonprofit arm of the farm, Green-Wave, Smith is sharing his model with others in order to create an army of sustainable seafood farmers around the world. Key to the model is reliance on shellfish, which can grow quickly, reproduce in spectacularly high numbers, and improve the habitat for other resident species.

Across the globe, new efforts to farm shrimp and fish more sustainably are also under way, with mixed results. One of the biggest impediments to sustainable aquaculture of fish and shrimp is the continued reliance on feed made from wild-caught species such as anchovies and herring. For every pound of salmon raised, an estimated *five pounds* of such forage fish were ground up and used as feed. We could feed a lot more people if we just ate those feed fish directly. These schooling silverfish are underutilized as food for people and currently overutilized as food for fish, pig, and chicken farms. The pressure to turn these fish into feed is pushing even these fast-growing, sexually vibrant species to their limits. To address this issue, private and public funding sources are pouring money into research efforts that are experimenting with alternative protein sources for feed, including plant proteins such as soy and flaxseed, insects, and the leftovers from fish-processing plants. In the meantime, the most sustainable aquaculture occurs when the product requires no feed inputs, such as is the case with shellfish, or can grow on feed that doesn't require fishmeal or fish oil inputs. Vegetarian and freshwater catfish and tilapia can meet these goals—as can shrimp, if they are farmed without destroying the local mangrove habitat. Often, this means the shrimp are raised in tank-based systems on land.

Finally, new efforts to farm some of the most depleted seafood fare may be the only chance for saving a species. Like captive breeding programs for endangered white rhinos, the Bodega Marine Laboratory at the University of California, Davis is leading the charge to save the critically endangered white abalone. After nearly a decade of failed attempts to get males and females to spawn in tanks, scientists caught a break in 2012. Dr. Kristin Aquilino, one of the lead researchers with the white abalone program, was there when it happened: "We had the abalone in separate buckets and had gone through all the procedures to try to get them to spawn." This, Aquilino told me, includes playing some Barry White to help set the mood. Adding a little mild peroxide to the seawater is the final touch—scientists think it's a possible mimic of the chemical cues abalone release in the wild. Then the scientists simply waited. "The females had released some eggs, and we just needed one of the males to go." Walking around anxiously from bucket to bucket, Aquilino was watching closely. And that's when she saw it, a tiny puff of milky white drifting out from the base of one of the male's shells. "If I hadn't been there with pipette in hand to suck it up, I would have missed it." The amount of sperm was minuscule, barely adequate to fertilize half the eggs, but it was enough. "It is the closest thing to a miracle I've ever experienced," sighed Aquilino. "From that small burst we made about twenty new animals. That was a really big deal. The next year, we made about a hundred and twenty, and the year after that, two thousand. This year, we've probably hit over five thousand. We keep increasing."

Though the researchers have a long way to go before they have the tens of thousands needed to potentially restore the population, the significant strides made in the past four years are reason for great hope—the first ray of light in what has been a dark history for abalone along North America's western shore. With continued success, we might just see white abalone back in the wild in the not-too-distant future.

GIVE THEM SOME PRIVACY: THE RISE OF
LARGE-SCALE MARINE PROTECTED AREAS

Distributing our demand, if it happens at a large scale, can certainly help species recover; but reversing declining ocean health on a global scale also requires that we build in some space where marine life can recover and thrive undisturbed. Just as we do on land, we need to set aside refuges that can help build up the capital and the resilience of ocean ecosystems.

The first national park in the United States, Yellowstone, was established in 1872; the first national marine sanctuary, at the site of the wreck of the USS *Monitor,* in 1972. Today, underwater marine protected areas (MPAs) continue to lag behind terrestrial initiatives: protected areas cover about 10 to 15 percent of all land, but only 2 to 3 percent of the ocean's area. And of that small amount, only about 1 percent of the ocean is fully protected in what are often called marine reserves—areas closed to all extractive activities, such as fishing or oil and gas operations.

For fish or sharks or squid or crabs on the hunt for a romantic getaway, marine reserves offer that perfect destination. But not all marine reserves are created equal. The size, amount of enforcement, and duration of protection all affect the success of a reserve in restoring marine species; historically, reserves have been too few, too small, and too poorly enforced to really help. But over the last decade, this has started to change. In 2004, the Great Barrier Reef Marine Park—which is nearly the size of Germany—expanded its fully protected area to one-third of the park. Since then, there has been an uptick in the number of countries designating large marine protected areas (many at least 100,000 square kilometers, which is just over 40,000 square miles) in their own territorial waters. The majority of these more recent designations are completely closed to fishing. Ranging from a park just over 58,000 square miles in Chile to the more than 490 million square miles in the Pacific Remote Islands Marine National Monument, these new reserves offer protection to formerly underserved species: large, highly mobile fishes and sharks. Many also include large tracts of diverse, neighboring habitats, such as sea grass beds, mangroves, and coral reefs, which allows for sex-segregated populations that occupy different ecosystems to all receive protection. They also guard the home ranges of species that may travel large distances, such as grouper, to move from home turfs to spawning sites.

In addition to those benefits, large reserves in particularly remote locations also offer another unique opportunity: the fish don't flee when we spy on them. This lack of fear, combined with the elevated (and more natural) abundances of species, allows for new discovery. The first time scientists observed male fish violently head-butting one another in dominance disputes is one example. A known behavior for many ungulates (such as deer, pigs, and goats), this kind of aggressive combat never had been seen before in a fish. But the displays put on by the giant bumphead parrotfish in 2011 left no room for doubt that fish can ram heads as well as any bighorn sheep.

The largest herbivores on coral reefs, bumphead parrotfish can reach five feet in length and weigh upward of 165 pounds—as much as a grown man. Severely overfished throughout their range, scientists thought that bumpheads used their blunt, flattened forehead to bash chunks of coral as part of their grazing activity. That was until they heard the jarring crashes and saw the repeated bouts of fish slamming into one another off Wake Atoll.

In other locations, the population is likely too small to warrant such aggressive competition. But on Wake Atoll, dense schools elicited combat. The males start by swimming in parallel, as if sizing each other up. After a few moments, they swim apart and then quickly turn and face one another, like gunslingers before a showdown. Then, they speed forward and crash heads, as many as four times in a row, before a defeated (and likely dizzy) male will concede and swim away. This behavior coincides with early-dawn spawning rushes, suggesting males may smash skulls to win territories within spawning areas.

Part of the Pacific Remote Islands Marine National Monument, which includes several islands and covers 490,000 square miles, Wake Atoll and other marine reserves like this hold enormous value as windows into the unadulterated dating and mating strategies of even some of the most conspicuous members of marine ecosystems.

KEEP IT SAFE:
SEX TECH FOR MARINE PROTECTION

If you are thinking about giant whale condoms, no, it's not *that* kind of protection. Instead, we're talking about high-tech gear that provides new ways to monitor and study our interactions with ocean life—in particular, the advent of affordable and accessible "spyware" that makes patrolling big, remote marine reserves possible and gives us a look at fishing activities far out to sea.

The nation of Palau includes eight major islands and around 250 islets with a total land area of just under a 180 square miles. Its territorial waters, however, include about 230,000 square miles—an area nearly the size of France. How can a country that small enforce fishing laws over an area that large? Up until a few months ago, the answer was, not very well. But today, thanks to the help of some satellites flying a few thousand miles above the sea's surface, they can.

Known as Project Eyes on the Seas, the effort is led by the Pew Charitable Trusts and combines satellite-based vessel-tracking information with targeted policy in order to help governments identify and prosecute illegal fishing within their waters. All vessels over a certain size are required to use an automatic identification system (AIS), which transmits signals containing a unique vessel identification number (like a VIN number on a car) and a GPS location. Pew's pilot "Virtual Watch Room" currently monitors activity around Chile's Easter Island and the national waters of Palau, helping these governments ensure protection of their vast marine resources even in remote locations.

In a complementary effort, the nonprofit SkyTruth is working with Google and the nonprofit Oceana to develop Global Fishing Watch, a web-based tool for monitoring fishing activity around the globe using the same AIS data as Pew, but in a different way. Currently, the tens of thousands of data points recorded by AIS every day are all fed into one giant cloud, which makes it difficult to differentiate fishing activity from just transit activity of fishing vessels to and from fishing grounds. SkyTruth is developing algorithms that can determine the kind of fishing taking place—bottom trawling versus long-lining, for example—based on the movements of fishing vessels. This open-access data will be available via an online public web portal, allowing viewers to help monitor the activities of fishing vessels around their favorite dive sites, off their home shores, or wherever they may be interested in exploring. By making this data open access, Global Fishing Watch aims to leverage the power of the public to identify best practices as well as those that are suspicious. It also hopes to compile information about how much fishing activity, and of what nature, is happening where and when around the world. Although these satellite technology solutions are nascent, they hold promise for changing the rules of the game.

From the stratosphere down to the subcellular, new tools are also allowing for better enforcement of fisheries regulations through the use of DNA testing. As is the case with any regulation, a rule is only effective if it can be enforced. In 2014, five new species of shark were added to Appendix II of the Convention on International Trade in Endangered Species (CITES), limiting international trade of these species and their parts. The recent listings brought the total number of sharks under Appendix II to eight species, including great whites and three hammerhead species. This international agreement is a major step forward in conservation of these threatened species, but it also presents an enforcement challenge.

Approximately 63 million sharks (and perhaps as many as 273 million) are harvested *every year,* primarily to supply the market for shark fin soup, a dish that serves as a status symbol in some Asian cultures. Shark finning to supply this market is one of the most wasteful and gruesome practices, with fishers slicing off the fins of animals on board and then throwing the still-alive bodies back overboard, where they bleed to death or drown, unable to swim without their fins.

In the initial stage of processing, shark fins are dried and shipped to Asia with the skins still on. Using newly developed shark-fin ID guides, trained customs officers can identify species from these fins and use DNA tests to confirm the species. The next processing step, however, removes the skin and uses chemicals to bleach the product, making species ID by sight impossible and damaging the DNA, rendering standard genetic tests unusable. However, a new technique, developed by a group of researchers, including the previously mentioned Demien Chapman of Stony Brook University and Kevin Feldheim of the Field Museum of Chicago, requires only partial sequences of DNA. So far, this "mini-DNA barcoding" test can successfully identify seven of the eight species at this later processed stage and even has some success with identifying species once they are in the soup. Similar DNA tests are being used to identify mislabeled seafood in restaurants and stores across the globe.

In addition, genetics also help map the routes of some of the ocean's biggest wanderers. For example, genetics, combined with photographs of a blue whale sampled in the Galápagos and off Chile, proved this one whale migrated a distance of approximately 3,200 miles, the longest distance of any Southern Hemisphere blue whale ever recorded. This information helps researchers identify critical habitats, including breeding grounds, for these long-ranging species. The data also helps frame efforts to establish networks of MPAs that can effectively protect these still-endangered populations.

Technology also can be used proactively to help whales achieve safe passage into and out of their favored habitats. Shipping traffic has increased greatly in recent years in Massachusetts Bay's Stellwagen Bank National Marine Sanctuary, increasing risk of ship strikes to those frisky and highly endangered northern right whales. Currently numbering around 490, the loss of any individual, especially an adult female, makes a big difference in the future survival of the species. Through a research-industry partnership, scientists at Cornell Lab of Ornithology's Bioacoustics Research Program and the Woods Hole Oceanographic Institution developed a series of "smart" buoys that can detect whale calls from up to five nautical miles away and then send warnings to ships approaching the region. Ship captains can then slow their vessels and post lookouts, reducing the risk of ship strikes. This initiative provides a model that could be replicated in other parts of the world where ship traffic into and out of breeding and feeding grounds poses risks to whales.

Listening stations also continue to help elucidate the dynamics behind Nassau grouper orgies as part of the Reef Environmental Education Foundation's Grouper Moon Project. These data show when the fish start to travel to the spawning site, where they come from, how long they stay, and where they go. In an effort to educate local communities about the importance of protecting this large spawning aggregation, the researchers work closely with local schools to educate children about this unique event in their home waters. Based in large part upon these efforts, in 2011 the Cayman Islands government extended the seasonal ban on fishing on Nassau grouper spawning sites—current and historical locations—for another eight years. Thus far, the ban seems to be working, with signs of recovery in the size of individual fish, new "teenage" fish showing up for their first spawns, and approximately double the number of fish since 2003 when the ban first went into effect.

Today, researchers are deploying more and more acoustical arrays along the coastlines of the world in order to help improve understanding of the migration patterns of species to and from spawning and feeding grounds. These sound stations are providing a whole new way to study species—by listening to where they go, and how (and if) they are communicating along the way.

GET IN THE ZONE: OCEAN MANAGEMENT
ON SCALES THAT COUNT

Of course, not every part of the ocean can be cordoned off as a marine reserve, nor does it need to be. Ocean zoning acknowledges the multiple uses of ocean resources and seeks to manage ocean areas as we do on land, with different zones permitting different activities. This includes consideration of important habitats for marine species, including breeding sites and nursery areas, as well as traditional fishing grounds, other extractive activities, and recreational and cultural uses. In the majority of locations around the globe, such holistic management of near and offshore ocean resources is lacking; fish are managed as single species, and often different government agencies regulate different activities with little coordination.

The Waitt Institute's Blue Halo Initiative is one example of how this traditionally fragmented approach can be replaced by a more comprehensive and effective one. Launched in 2012, this initiative works with island nations to identify and accomplish their goals for ocean use and management across all their territorial waters, up to three nautical miles from shore. A nascent effort, the initiative has successfully led to the signing of ocean management legislation in Barbuda, and has recently launched in Curacao and Montserrat.

Smaller reserves can have their success, too, especially those that have been enforced for several years. Cabo Pulmo National Marine Park is one such success story. Located in what Jacques Cousteau once called "the Aquarium of the World," fishers living on the southeastern tip of Mexico's Baja Peninsula noticed that after decades of overfishing, the ancient reefs they had always relied upon for sustenance were empty. In an effort to save the ecosystem from complete collapse, local fishers organized and petitioned the Mexican government to create a marine protected area, including a large proportion of no-take zones.

It was a bold move that depended upon engagement and enforcement by the local community. It took about five years before signs of recovery occurred, but between 1999 and 2009, the marine reserve gained an astounding 463 percent more fish than nearby waters. From small to large, the fish came back in droves, including the return of spectacular spawning aggregations of jacks, grouper, and snapper. This kind of targeted approach to protect specific areas of high value and vulnerability could also be used to protect seamounts—especially if combined with the more advanced monitoring capacities of new technologies. Full protection is not necessarily always needed, but for sensitive habitats under increasing pressure (technologies now exist that could open seamounts to mining activity, for example), a combination of closed and carefully controlled access is needed.

When combined with other fishery or coastal development regulations, small reserves strategically placed—and properly enforced—can lead to big impacts. Back in the late 1990s, female gag grouper (yes, they really are called gag grouper) faced the challenging prospect of trying to spawn in aggregations where fewer than 5 percent of the individuals were males. A protogynous hermaphrodite, the gag grouper is born female and transitions to male at about ten or eleven years old. Over time, fishers had preferentially removed most of the large, sexually mature males and also had removed many of the larger females that would have eventually become male. With so few males, there was likely not enough sperm to go around to fertilize all the females' eggs. Female gag grouper ripe with eggs were showing up at the party but many never spawned. To address the declines and skewed sex ratio, managers created seasonal closures during the months of spawning, which allowed the adults to focus on sex rather than being distracted by tasty bait. In addition, small areas known to harbor large, sexually mature males were closed to fishing permanently. These measures, combined with annual catch limits, have resulted in a recovery of the gag grouper species in the South Atlantic and Gulf of Mexico in recent years.

GIVE 'EM YOUR BUSINESS: SUPPORT ALTERNATIVES TO FISHING

Curtailing fishing activity is easier said than done, especially for spawning aggregations, because many fishers depend upon these concentrated catches as a major source of their yearly income. Finding alternative livelihood support is critical to creating effective policy, and helps incentivize compliance with regulations. One potential approach is to turn fishers into paid guides

for voyeuristic divers wanting to sneak a peek at the action on the spawning grounds. In this way, a single fish spawning year after year can bring a
fisherman many more dollars than it would if hooked and sold once. Both
Belize and Cuba have experimented with this approach, retraining fishers
for the tourism or recreational fishing industry with some success.

Similarly, the rise of the underwater ecotourism industry provides
opportunities for businesses to profit from protection of marine species,
not just extraction of them. In island nations from the Bahamas to the
Maldives, resident shark populations now drive a dive industry that gen-
erates millions of dollars per year. In Palau—the first country to ban commercial fishing of sharks in all its territorial waters—a recent economic
evaluation estimated a single resident shark would generate nearly US$2
million over its lifetime compared to a couple hundred dollars finned and
sold as meat. Altogether, living sharks are likely bring in at least 8 percent
of Palau's GDP. From manta rays to shark dives to whale watching, such
studies are showing that many marine species are worth far more alive
than dead. This kind of reassessment of the value of marine species has
contributed to the growing trend of establishing shark sanctuaries around
the globe. In the last few years, nearly a dozen countries have joined Palau in banning commercial shark fishing in their waters. When choosing
a destination for vacation, especially for all you recreational divers out
there, consider spending your tourism dollars in countries that are working to preserve and protect their marine resources, rather than those that
continue to serve shark fin soup.

Ecotourism isn't feasible in every location, and in those places, other
forms of compensation and livelihood must be developed. In an example
of "ecopreneurship," at the other end of the food chain, social entrepre-
neur Wayan Patut, from Serangan Island just south of Bali, Indonesia,
created a new business model for helping turn fishers into coral farmers.
After a big resort development project devastated the local reefs, Patut engaged the children of local fishers to create small coral farms, growing
their corals around the reefs where their parents were now using destruc-
tive fishing practices, such as dynamite, to catch fish. It was a clever—and

intentional—tactic. Moved by the children's efforts to restore the habitat and remiss to harm the children's coral gardens by dynamiting for fish, the fishers slowly began to learn the farming techniques themselves. The co-operative that Patut created now grows corals to sell to the aquarium trade, while also out-planting some for wild reefs. Income generated from the sale of coral fragments, as well as reef fish for the aquarium trade, along with snorkel tours of the local farms, helps provide income to fishers; they now make a living helping build up, rather than blow up, coral reefs.

The science behind coral farming is developing at a fast pace, with dozens of nurseries around the globe actively growing and out-planting coral fragments onto damaged and recovering reef sites in the hopes of boosting local population numbers. As research improves coral propaga-tion techniques, the cost of growing corals drops and farming becomes far more user-friendly. For example, Mote Marine Lab has already discovered how to accelerate growth rates in the Florida Keys based on the shape and size of the fragments, while researchers in Australia are experimenting with growing coral species that may be more tolerant to climate change impacts. This surge of coral farming has spawned the development of an active "volunteerism" effort, where visitors to reefs can take part in growth and transplant programs. And it has opened the door for eco-minded entrepre-neurs to create ocean-friendly businesses.

GO AU NATUREL: ADDRESSING POLLUTION
AND COASTAL DEVELOPMENT

Sex happens best when it happens naturally. For marine species, this means having habitats that are free of the plastic and pollution that causes un-natural sex change or otherwise disrupts successful reproduction. Even in the sea, everybody likes cleans sheets.

To date, increased levels of endocrine-disrupting chemicals are associ-ated with delayed spawning of English sole in Puget Sound, feminization of male mullets along the Basque coast, sterility in ringed seals, demascu-linization of males in some crustaceans, masculinization of some females (the penis-sprouting in snails), decreased numbers of eggs in cod, and of course the strange intersex amphipods Dr. Ford has been finding along the Scottish coast. Thirty years after their ban, high polychlorinated biphenyl loads continue to cause reproductive failure in harbor porpoises, while hormones used in the cattle industry appear to turn solicitous, flirtatious male guppies into covert rapists. There are also impacts on communication: chemical pollutants from runoff, leaking undersea oil pipes, and of course major oil spills also can interfere with mating success by masking the chemical signals constantly wafting through the sea.

As for the prevalence of plastic, the most recent study estimates between eleven billion and twenty-six billion pounds of it enter the ocean *every year* from land. Reducing the input of pollution into the sea is often a matter of changing policy and behavior. This can happen at global and local scales to great effect. The worldwide movement to ban plastic bags is an example of a local effort that has gone global, with complete bans in countries as diverse as Italy and the Ivory Coast, while citywide bans occur in countries from the United States to Australia, Pakistan, and India. Together, these efforts reduce the amount of plastic bags headed to landfills and the oceans on a large scale. Similar efforts to pass legislation banning the use of plastic microbeads, used as abrasives in cosmetics, toothpastes, and soaps, are also under way in the United States at both the state and federal levels. These beads are so small they tend to pass straight through municipal water treatment facilities and head out to sea, where they can accumulate toxins or fill up the bellies of sea life, causing malnutrition.

Recent data indicates that more than 80 percent of plastic pollution comes from about two dozen middle-income, rapidly developing countries that lack good waste management systems. This is actually good news, as it means a concentrated effort to improve facilities in these target countries could lead to a major decrease in future ocean pollution.

In terms of the pollution that is already in the sea, straining out the really small particles is nearly impossible—anything that can capture the small stuff also takes with it all the important plankton the food web depends on. This presents one of the biggest challenges for ocean management to date. On the other hand, however, we are making progress with the big chunks.

In Hawaii, a novel program called Nets to Energy transforms ocean trash into an energy source for the islands and serves as a fantastic model for other regions. NOAA's marine debris program, along with other non-profits, private industry, and fishers, collects derelict fishing nets that snag on corals, endanger wildlife, and clog the shorelines. These nets are then ground into tiny pieces and burned at a waste-to-energy facility, producing steam that powers turbines. So far, over eight hundred tons of nets have been converted into enough energy to power three hundred homes for a year.

As the saying goes, many hands make light work. In 2014, more than a half-million people, comprising a million hands, picked up more than sixteen million pounds of trash in ninety-one countries as part of the Ocean Conservancy's International Coastal Cleanup campaign. This is a three-decade-long volunteer effort to pick up and record trash on beaches around the world. Besides cleaning the coasts, this program also collects data on every piece of garbage, helping scientists study how trash moves across the seas, and what we can do to minimize this impact.

Global action to ban harmful chemicals is not beyond reach, either. Remember the penis-sprouting female snails? The cause of their distress was tributylin (TBT), a ubiquitous and noxious chemical found in the paint used on the bottoms of boats. A worldwide ban on TBT went into effect in 2008, despite the fact that the shipping industry relied heavily on this paint for antifouling purposes. Recent reports indicate that the ban appears to be working, with penis growing on a downward trend in female snail populations. Likewise, stronger regulations on coal emissions in the United States has significantly reduced mercury input into the air and sea. One result has been a faster-than-expected decline in mercury content in top predator bluefish off the Eastern Seaboard.

There are other ways we can promote more natural conditions for sea life, especially along the coasts. A surge of projects to rebuild oyster reefs is one such example. From specially designed concrete blocks to bags packed with old oyster shell, new techniques to construct starter reefs help lift small oysters off the silty seafloor and provide a more attractive habitat on which wild oyster larvae can settle and grow. People with waterfront properties are joining "oyster gardening" efforts in which they hang bags of juvenile oysters off their docks, providing a protective way to grow young oysters to larger sizes, and then placing mature oysters out on restoration sites. Other ways folks are giving back to help oysters include shell recycling programs, such as those run by the Oyster Recovery Partnership in Maryland and the Coalition to Restore Coastal Louisiana. Diners slurping back oysters at participating restaurants can rest assured valuable shells head back to the sea (rather than to the dump), where they can help build up reefs and attract new wild oysters to settle. In Maryland, businesses that recycle their shells can even collect a tax credit. In fact, Maryland now boasts claim to the largest restored oyster reef in the world, at 330 acres. The approximately one billion oysters making up this reef were carefully cultivated in the lab, settled onto shells saved from the landfill, grown until they reached a large enough size, and then released back into the brine. They are part of an enormous effort to use oysters to clean up the pollution that is plaguing the once bountiful Chesapeake Bay.

A growing awareness of the multiple benefits of living barriers, as opposed to manmade concrete bulwarks, has led to more local municipalities, from Texas to New York, considering oyster reef restoration as a smart investment—one that returns not only coastal protection, but also improved water quality, fisheries, and recreational value.

CREATE THE CLIMATE FOR SEX

Preventing the worst-case scenario of climate change requires global collaboration at the highest levels of government—never easy to achieve. Yet there is reason to hope. The oceans are proving more resilient than initially thought. Make no mistake: if our use of fossil fuels remains unchecked, the results will be dire—coral reefs as we know them will cease to exist, for example. But the adaptive capabilities of marine life are showing there is still time for us to course correct. For example, warmer temperatures are likely to skew sex ratios of offspring in many species. Yet in recent experiments on reef fish, parents resolved this imbalance within two generations—fish born in the warmer waters adjusted to produce more equal male-female offspring ratios. These studies indicate that the same flexibility in sexual determination that may leave species vulnerable to the impacts of climate change may work to buy us some time as species are able to adapt to minor shifts without resulting in irreversible damage.

MPAs can also help populations bounce back from potential climate-induced effects. Off the Pacific coast of Baja California, abalone populations inside fully protected reserves did not suffer declines as significant as populations outside the reserve following two severe low-oxygen events—a phenomenon expected to become more frequent with changing climate. Although all abalone were engulfed by the dead zone, the much higher starting densities and larger size (hello, BOFFFFs!) of adult abalones inside the reserves provided a buffer. Enough abalone survived in close proximity to each other for successful spawning and recruitment, resulting in faster recoveries than outside the reserve. The reproduction rate inside the reserve was high enough to also boost settlement of juveniles outside the reserve boundary, in what is known as a "spillover" effect.

Similarly, in the Caribbean, areas of well-managed reefs, especially those with healthy populations of grazing parrotfish, show resistance to negative impacts from warming temperatures. These findings indicate that local actions can make a difference in the long-term survival of these ecosystems.

If we can act to prevent the most extreme climate change scenarios, ocean ecosystems still stand a chance. Unlike on land, extinctions have been rare in the sea, and when placed under appropriate management, even some of the most overfished species—such as humpback whales or West Coast rockfish or North Atlantic swordfish—have shown it is possible to come back from the brink.

THE SEA IS ONE
SEXY BEAST

THE STORY OF SEX IN THE SEA IS KINKY AND CON-
cerning; a tale of the magnificent diversity of ways that life begets life in
the ocean but also the myriad challenges to management that such diversity presents. The good news is that Nature is on our side here.

The urge to reproduce is so strong, it drives some of the most extreme
behaviors on the planet. It will propel whales and sea turtles to crisscross
ocean basins; compel a male to part with his penis; persuade a female to
abandon her aquatic home for shoreline bondage. It is a force to be reckoned with, and it is where I find inspiration when faced with a seemingly
endless list of threats to ocean life.

As I write this, for example, the warm summer night air reminds me
that beneath the surface in the heart of the Caribbean Sea, one of the
most endangered ecosystems on Earth is about to erupt in possibility. This
very week, millions of coral polyps are putting the finishing touches on
their egg-sperm bundles, wrapping a year's worth of stored-up energy in
bright pink, spherical packages. Now the bundles begin their silent march
upward from deep inside the coral body toward the tiny opening at the
mouth. Slowly and steadily, and still held within the coral polyp, they are
already rising toward the surface.

Very soon, a few days after the full moon in August, the boulder star corals will let loose—millions upon millions of them in a perfect, ritualistic synchrony of sex that has gone on for millennia and still continues year after year. Knowing this, having seen it firsthand, is deeply reassuring. There is something sacred in witnessing the repetition of Nature's yearly cycles with such elegant exactitude, even in the face of such extraordinary change. Our opportunity and our responsibility lie in ensuring this potential continues to be realized. All the raw material is still there. Despite all the threats, for now, the corals spawn on. If ever there was something to be hopeful about, it is this.

ACKNOWLEDGMENTS

JUST LIKE SOME OF THOSE WHALE VAGINAS, MY journey through *Sex in the Sea* has been long and winding. There are many thanks to give.

The stories nested between these covers reflect the expertise of many scientists who graciously gave their time to speak with me, shared publications, sent thoughts via e-mail from remote field stations, and contributed images for this book. It is their dedication that builds the data, their tenacity that moves the field forward, their insights that provide greater understanding and the raw material for forging change. For your time, your sense of humor, your candor, and your patience, I gratefully thank: Octavio Aburto-Oropeza, Kristin Aquilino, Jelle Atema, Andre Boustany, Demian Chapman, Phillip Clapham, Diane Cowan, Ted Cranford, Peter Dutton, Brad Erisman, Kevin Feldheim, Nicole Fogarty, Alex Ford, Peter Franks, Jim Gelsleichter, Matthew Grober, Dean Grubbs, Kristin Gruenthal, Juliana Harding, Phil Hastings, John Hildebrand, Henk-Jan Hoving, Ayana Johnson, Peter Klimley, Nancy Knowlton, Don Levitan, Mark Luckenbach, Elizabeth Madin, Kristin Marhaver, Bruce Mate, Steve Midway, Paul Olin, Dara Orbach, Steven Ramm, Victor Restrepo, Greg Rouse, Yvonne Sadovy, Brice Semmens, David Siveter, Dan Spencer, Josh Stewart, Bob Warner, and Jeanette Yen.

A special thanks to Sarah Mesnick, with whom I have enjoyed ongoing conversations about the ever-evolving landscape of sexual strategy in the sea since 2005. Thank you for your mentorship and especially for your contagious enthusiasm for the sex lives of beasts from blennies to belugas. I am grateful to John Amos, founder of SkyTruth, for bringing me up to speed on the high-flying capabilities of satellite technology, and to Bren Smith and Barton Seaver for multiple conversations over the years regarding how to build solutions to the challenges of sustainable seafood.

To Dr. Samuel "Doc" Gruber, a long-overdue thank-you for giving a burned-out teen the opportunity of a lifetime and cementing my love of sharks. A heartfelt thanks to my PhD adviser, Dr. Jeremy B. C. Jackson, who—as only a true Renaissance man and renegade scientist would—took a chance on a history of science major. Thank you for giving me the opportunity to become a marine biologist, and especially for teaching me to look for the story in the data. To Dr. Carl Safina, thank you for your early mentorship in the craft of science communication.

Thank you to my agent, Michelle Tessler, for guiding me through this new and foreign world and finding my manuscript a home. And thanks to my editor, Elisabeth Dyssegaard, and the entire staff at St. Martin's Press for their support and patience in handling the questions and concerns of a newbie author.

It's not easy to capture the characters of *Sex in the Sea* in the act. Through extraordinary patience and talent, the following photographers managed to reveal (or at least provide hints) at what may lie beneath the veiled surface. Most of these folks have beautiful websites and I encourage the reader to explore them—these galleries are especially soothing when one is feeling a little landlocked. To each of you, I am honored to be able to feature your art in this book and appreciate all the generosity of spirit you have shown in helping to make this collection of images possible. With warmest thanks to Octavio Aburto-Oropeza, Kristin Aquilino and the Bodega Marine Laboratory White Abalone Captive Breeding Program, Tim Calver, Bryce Groark, Jillian Morris, Raphael Ritson-Williams, Christy Semmens and the Reef Education and Environment Foundation, and Klaus Steifel.

I want to express deepest gratitude to my mentor and boss, Cheryl Dahle, who supported this effort from the start. Your leadership in building an organization where each individual can truly thrive is a rare gift, and one that I am grateful for and inspired by every day. Additional appreciation as well for your always insightful edits, sense of humor, and reality checks when I needed them most. To all my colleagues at Future of Fish / Flip Labs, thank you for your patience and accommodation—your indefatigable commitment to creating positive change in this world is simply invigorating, and I look forward to diving back into the fray with you. A special thanks to Dr. Colleen Howell for keeping the research department ship afloat and providing such insightful feedback, again and again. Your grace and selfless spirit are such a light in this world.

To the oh-so-talented Missy Chimovitz with Antigravity Studios: your witty, whimsical, and wonderful illustrations have brought these chapters to life beyond my wildest dreams—and let me tell you, that's saying something. Thank you for your commitment to supporting my vision while bringing in your own marvelous sense of humor and creativity. Your friendship and skills have been extraordinary gifts in this process, and I am so grateful for both.

Completely aware of the terrible pun, I must note that this book has been a labor of love nearly a decade in the making. It has weathered seven moves, three job changes, and the early years of parenthood. The capacity to follow through on an idea for as long as this was only possible because of the support I received from my family and friends. For patience with dinner conversations dominated by musing over reproductive strategies of marine life, I humbly thank the Andersons and the Cambria Friday Night crew. For always colorful feedback on proposed euphemisms for sex, my gratitude to Lil' Mamas. Overwhelming gratitude to the Mag 7 for the constant cheerleading, advice on how to balance humor with seriousness of subject, and the best dirty senses of humor around—I couldn't have done it without you ladies. To all those who provided edits and thoughts on various stages of the text, I am grateful for your kindness, corrections, and encouragement, especially Gen Marvin, Sarah Kalloch, Ann Johnson, Frances Lloyd, and Rebecca Marsick.

For the gift of quiet writing space, I am grateful to Maxine and Chuck Lobel, Corky and Olga Tamboer, and my folks for lending me their homes to write in; I am also indebted to the owners of numerous coffee shops spanning Hawaii, California, Connecticut, and Colorado for countless hours of camping out. For an early kick in the ass to get writing already and for valuable research assistance, I thank Jane Hirsh.

Enormous love and appreciation to my mom and dad, who quickly shrugged off the inherent awkwardness that comes with discussions of sex between parent and child and dove into countless conversations about the fascinating world of underwater penises. A special thanks to my mom for plowing through numerous early and later drafts. A heartfelt thanks to my sister and brother-in-law for the constructive criticism only siblings could provide.

And to Maddox, thank you for sharing your mama with these words on a page for so long—in ten years you may be extremely embarrassed, but a few more years after that, I hope you will understand how so much of this is for you.

Last, but most of all, my endless gratitude and love for the person who made it all possible. Thank you, Steve, for traveling through the depths and pulling me back to the sunlit shallows when I needed it most. You make a life on land a beautiful thing.

NOTES

INTRODUCTION: GETTING YOUR FINS WET

inside his giantess mate's kidney: A species of deep-sea echiuran worm called *Bonellia viridis* is one example of this extreme romantic pairing.

Nearly three billion people rely on fish as major source of protein: "Major source" equals about 20 percent of total animal protein consumption. See "The State of World Fisheries and Aquaculture: Opportunities and Challenges," United Nation Food and Agriculture Organization, 2014, http://www.fao.org/3/a-i3720e.pdf.

CHAPTER ONE: THE QUEST

99 percent of the habitable space on the planet: Although about 70 percent of the Earth's surface consists of water, in terms of volume, the ocean provides 99 percent of the habitable space on the planet. "Living Ocean," NASA Science Earth, October 5, 2015, http://science.nasa.gov/earth-science/oceanography/living-ocean/.

Many of the eleven thousand or so species: For up-to-date references on current copepod studies and counts, see the World of Copepods website: http://www.marinespecies.org/copepoda; also A. G. Humes, "How many copepods?" *Hydrobiologia* 292–293 (1) (1994): 1–7.

when you are as small as a copepod: My primary source for the detailed description of what life is like for copepods in the sea and how they find their mates is Dr. Jeanette Yen, professor of biology, Georgia Institute of Technology, in multiple interviews that took place January and February 2015.

up to one hundred body lengths away: Michael H. Doall, Sean P. Colin, J. Rudi Strickler, and Jeannette Yen, "Locating a mate in 3D: the case of *Temora longicornis,*" *Royal Society* 353 (1998): 681–689.

thin section of water: Dr. Peter Franks at Scripps Institution of Oceanography first introduced me to this concept of "singles bars" for copepods way back during my graduate studies, and reminded me of their existence during our later interview. Yen then similarly described the phenomena, referring to them as "copepod discotheques."

may have trouble finding these trusted singles bars: Yen noted that shifts in the location and stability of boundary layers are one likely outcome of warming surface conditions and increased storms.

copepod courtship dance: J. Titelman et al., "Copepod mating: chance or choice?" *Journal of Plankton Research* 29(12) (2007): 1023–1030, doi:10.1093/plankt/fbm076.

pursue virgin females: J. Heuschele and T. Kiørboe, "The smell of virgins: mating status of females affects male swimming behaviour in *Oithona davisae,*" *Journal of Plankton Research* 34(11) (2012): 929–935.

"rejection dance": G. Dur, et al., "Mating and mate choice in Pseudodiaptomus annandalei (Copepoda: Calanoida)," *Journal of Experimental Marine Biology and Ecology* 402(1–2) (2011): 1–11, doi:10.1016/j.jembe.2011.02.039.

are not perfectly symmetrical: E. A. Ershova and K. N. Kosobokova, "Morphology of genital system and reproductive biology of the arctic calanoid copepod *Metridia longa*," *Biology Bulletin* 39(8) (2012): 676–683. doi:10.1134/s1062359012080043.

only last for two or three days each year: Detailed description of Nassau grouper spawning aggregation formation was provided by Dr. Brice Semmens in an interview by the author on June 25, 2015. For more information and a video of the a spawning event, see his talk, "Grouper Moon: saving one of the last great populations of an endangered Caribbean reef fish," Jeffery B. Graham Perspectives on Ocean Science lecture at the Birch Aquarium at Scripps, February 2, 2012, http://www.uctv.tv/shows/Grouper-Moon-Saving-One-of-the -Last-Great-Populations-of-an-Endangered-Caribbean-Reef-Fish-25958.

The Grouper Moon Project: This is an initiative of Reef Environmental Education Foundation in partnership with the government of the Cayman Islands, Scripps Institution of Oceanography, and Oregon State University.

display bold, dramatic tones: Leslie Whaylen et al., "Observations of a Nassau grouper, *Epinephelus striatus,* spawning aggregation site in Little Cayman, Cayman Islands, including multi-species spawning information," *Environmental Biology of Fishes* 70 (2004): 305–313.

several theories about what makes for a good spawning site: Briefly reviewed in M. Russell et al. "Status Report: World's Fish Aggregations 2014," Science and Conservation of Fish Aggregations, California, USA, International Coral Reef Initiative, 2014, http://www.scrfa .org/images/stories/pdf/Status_Report_Worlds_Fish_Aggregations_2014.pdf.

every single adult Nassau grouper: Semmens, interview by author, June 25, 2015.

Maybe it just takes longer: Semmens notes that part of the focus of the Grouper Moon Project is to start to understand the more intimate dynamics of these smaller spawning aggregations.

tuned to the precise frequency: J. M. Gardiner et al., "Sensory Physiology and Behavior of Elasmobranchs," in *Biology of Sharks and Their Relatives,* ed. J. C. Carrier, J. A. Musick, and M. R. Heithaus, 2nd ed. (Boca Raton, FL: CRC Press, 2012), 349–402.

host a diversity of life: Telmo Morato et al., "Seamounts are hotspots of pelagic biodiversity in the open ocean," *PNAS* 107(21) (2010): 9707–9711.

sharks descend upon: Dr. Peter Klimley, adjunct professor at the University of California, Davis, has studied the behavior of hammerhead sharks around seamounts in the Gulf of California for over twenty years and provided insight into their dating games in an interview by the author, June 29, 2015.

it's a slow motion free fall: The BBC caught footage of the free-fall mating of two hammerhead sharks, which can be viewed on the not-for-profit ARKive's Scalloped hammerhead species webpage, http://www.arkive.org/scalloped-hammerhead/sphyrna-lewini/video-09a .html.

the way sharks navigate: Peter A Klimley, John E. Richert, and Salvador J. Jorgensen, "The home of the blue water fish," *American Scientist* 93(1) (January/February 2005): 42–49.

have become popular fishing sites: Due to their remote and deep locations, seamounts are one of the least well-studied ecosystems. *However,* what we do know is that fisheries on seamounts tend to be boom and bust, with rapid depletion of targeted species and high levels of habitat destruction associated with fishing methods (mostly bottom trawling). For more information on seamounts, their ecology, and exploitation, see Tony J. Pitcher et al., eds., *Seamounts, Ecology, Fisheries, and Conservation* (Oxford: Blackwell Science, 2008), 552. For anticipated impacts from mining, see T. Schlacher, et al., "Seamount benthos in a cobalt-rich crust region of the central Pacific: conservation challenges for future seabed mining," *Diversity and Distributions* (2013): 1–12.

That's a salmon's sex life: For details on the spawning tendencies of salmon, I relied on the expertise of Dan Spencer, fisheries consultant. E-mail exchange with author, September 15, 2015.

the same beach: In general, northern elephant seals appear to return to natal rookeries year after year; however, immigration to new beach areas can occur, especially as colonies recover and the beach gets more crowded. B. S. Stewart et al., "History and Present Status of the Northern Elephant Seal Population," in *Elephant Seals: Population Ecology, Behavior, and Physiology,* ed. B. J. Le Boeuf and R. M. Laws (Berkeley: University of California Press, 1994), 29–48.

eight months at sea: The Friends of the Elephant Seal's website, http://www.elephantseal. org, provides a great overview of elephant seal biology, ecology, and behavior.

one satellite-tagged female: Daniel P. Costa, Greg A. Breed, and Patrick W. Robinson, "New insights into pelagic migrations: Implications for ecology and conservation," *Annual Review of Ecology, Evolution, and Systematics* 43(1) (2012): 73–96, doi:10.1146/ annurev-ecolsys-102710-145045.

most will die virgins: Sarah L. Mesnick and Katherine Ralls, "Mating Systems," in *Encyclopedia of Marine Mammals,* ed. W.F. Perrin, H. G. M. Thewissen, and B. Würsig (San Diego, CA: Academic Press, 2008), 712–719.

sea turtles are another species: Information about the mating behavior of sea turtles was provided by Dr. Peter Dutton, program leader, Marine Turtle Genetics Program at NOAA's Southwest Fisheries Science Center, in a conversation with the author on April 10, 2015. More specific details regarding sea turtle's cuing into magnetic fields came from Costa et al., "New insights into pelagic migrations," 86.

a Wildcoast ad campaign: Dutton first alerted me to this "sexy" approach to conservation. More information about the reach and scope can be found on the Wildcoast website at http://www.wildcoast.net/who-we-are/heroes/12-dorsimar.

temperature of the sand determines the sex of the hatchling: The influence of external conditions on sex is known as "environmental sex determination."

six sharks tagged as babies: K. A. Feldheim et al., "Two decades of genetic profiling yields first evidence of natal philopatry and long-term fidelity to parturition sites in sharks," *Mol Ecol* 23(1) (2014): 110–117, doi:10.1111/mec.12583.

from koalas to humans: A recent study by Benjamin Charlton and colleagues revealed that male koalas have a novel vocal organ in their throat that allows the males to produce low-frequency "bellows" on the order of what is expected from a much larger elephant. B Charlton et al., "Koalas use a novel vocal organ to produce unusually low-pitched mating calls," *Current Biology* 23(23) (2013): R1035–R1036.

still prefer guys with a deep voice: Y. Xu et al., "Human vocal attractiveness as signaled by body size projection," *PLoS ONE* 8(4) (2013): e62397.

The theory is a work in progress: The acoustics behind this fascinating study were explained to me when I interviewed Dr. Sarah Mesnick of NOAA's Southwest Fisheries Science Center and through e-mail correspondence with Dr. John Hildebrand, professor at Scripps Institution of Oceanography. Both are coauthors of the study that showed this downward shift in blue whale song frequency, published here: Mark A. McDonald, John A. Hildebrand, and Sarah Mesnick, "Worldwide decline in tonal frequencies of blue whale songs," *Endangered Species Research* 9 (2009): 13–21, doi:10.3354/esr00217.

Blue whales escaped: A brief summary of the history of blue whale hunting was provided to me in correspondence from John Hildebrand and in my interview of Dr. Phillip Clapham of NOAA's National Marine Mammal Lab's Cetacean Program on February 9, 2015.

an estimated 380,000 blue whales were killed: Using newly found data from Russia's historic whaling industry, researchers now estimate a total of nearly three million whales were killed during the twentieth century. Robert C. Rocha Jr., Phillip J. Clapham, and Yulia Ivashchenko, "Emptying the oceans: A summary of industrial whaling catches in the 20th century," *Marine Fisheries Review* 76(4) (2015): 37–48, doi:10.7755/mfr.76.4.3.

males lead a solitary life: Background on sperm whale mating strategies was provided by Dr. Sarah Mesnick during our interview on May 5, 2014.

commandeering CT scanners: Dr. Ted Cranford, adjunct assistant professor at San Diego State University, provides images and a 3D video depicting the internal structure of the sperm whale head on his website at www.spermwhale.org. Details behind the sound production capacity of sperm whales and its potential function were provided by Cranford during two telephone interviews with the author on February 4, 2015, and July 27, 2015.

cascading series: A great place to experience this kind of effect for yourself, according to Cranford, is while standing under the large dome at the Palace of Fine Arts in San Francisco (other large domes above concrete floors with lots of reflective tile or stone also will likely do). Give a loud clap and the sound goes up, bounces off the curved dome, and is focused back down to you in a pattern reminiscent of the sperm whale's boom.

largest on the planet: Ted W. Cranford, "The sperm whale's nose: Sexual selection on a grand scale?," *Marine Mammal Science* 15(4) (1999): 1133–1157.

evolved to prefer: Sarah Mesnick proposes that behavioral and socially mediated responses to mate selection could lead to enhanced vulnerability in some species over others. This idea requires further testing, but is discussed in more detail in chapter 5, "Inner Chambers," and in an article she coauthored with Paul Wade and Randall Reeves, "Social and behavioural factors in cetacean responses to overexploitation: Are Odontocetes less 'resilient' than Mysticetes?" *Journal of Marine Biology* (2012), doi:10.1155/2012/567276.

declining numbers alone: Selective removal of the largest males likely led to drops in pregnancy rates in female sperm whales off the Galápagos for more than ten years after the cessation of hunting. See Hal Whitehead, J. Cristal, and S. Dufault, "Past and distant whaling and the rapid decline of sperm whales off the Galápagos Islands," *Conservation Biology* 11(6) (1997): 1387–1396.

whale off Newfoundland may be communicating with: Dr. Christopher Clark of Cornell University, in an interview regarding the expanded "acoustical landscape" of cetaceans, notes, "So if I am a whale off Newfoundland, I can hear a whale off Bermuda." David Brand, "Secrets of whales' long-distance songs are being unveiled by U.S. Navy's undersea microphones—but sound pollution threatens," *Cornell Chronicle,* February 2005.

The oceans are far noisier: C. W. Clark et al., "Acoustic masking in marine ecosystems: Intuitions, analysis, and implication," *Mar Ecol Prog Ser* 395 (2009): 201–222; L. T. Hatch et al., "Quantifying loss of acoustic communication space for right whales in and around a U.S. National Marine Sanctuary," *Conservation Biology* 26(6) (2012): 983–994.

woke up half the neighborhood: This bizarre phenomenon was covered by David Moye in his *Huffington Post* piece "Strange Hum Rocks Seattle: Are Mating Fish the Cause?" September 7, 2012, http://www.huffingtonpost.com/2012/09/07/strange-hum-seattle-mating-fish_n_1865142.html.

hormones enhance the female fish's ability to hear: J. A. Sisneros, "Adaptive hearing in the vocal plainfin midshipman fish: Getting in tune for the breeding season and implications for acoustic communication," *Integr Zool* 4(1) (2009): 33–42, doi:10.1111/j.1749-4877.2008.00133.x.

extraordinary flights: The following section is based on my telephone interview of Joshua Stewart of the Gulf of California Marine Program at Scripps Institution of Oceanography about his ongoing research to study the behavior and ecology of these fascinating aggregations of rays. July, 7, 2015. For beautiful footage of these rays in flight, check out BBC Earth at http://www.bbc.com/earth/story/20150512-watch-these-giant-rays-fly.

CHAPTER TWO: LURING A LOVER

In the world of lobster sex: I was immersed in the world of lobster sex by Dr. Jelle Atema of Boston University and Dr. Diane Cowan, founder of the Lobster Conservancy, during telephone interviews and while reading their numerous publications. For an in-depth look at the lives of lobster beyond just their sex habits, I highly recommend Trevor Corson's brilliant *The Secret Life of Lobsters* (New York: HarperCollins, 2005), which deftly explores the intimate relationship between this ancient crustacean and the people who fish and study them for a living.

they use their antennules: Antennules are the smaller of two sets of antennae sported by lobsters, with each antennule covered in many fine hairs that lead down to the olfactory receptors. Jelle Atema, conversation with the author, August 5, 2015.

remember the smell: M. E Johnson and J. Atema, "The olfactory pathway for individual recognition in the American lobster *Homarus americanus,*" *J Exp Biol* 208(pt. 15) (2005): 2865–2872, doi:10.1242/jeb.01707.

keep him entranced: The female will continually release her "potion" throughout her stay. The constant dosing helps subdue the male so that he allows the female to stay in his den for this extended period.

twenty-five million years: According to a recent study, the origin of the first lobster-like decapod (Polychelida) was estimated in the Devonian (ca. 409–372 million years ago), and the rise of modern day *Homarus* lobsters occurred sometime between about 25 and 110 million years ago, depending on the method applied. See Heather D. Bracken-Grissom et al., "The emergence of the lobsters: Phylogenetic relationships, morphological evolution and divergence time comparisons of an ancient group (Decapoda: Achelata, Astacidea, Glypheidea, Polychelida)," *Society of Systematic Biologist* 63(4) (2014): 457–479, especially figures 3 and 4, doi:10.1093/sysbio/syu008.

strong current: To breathe, lobsters pump water into their gill cavity, located near where their front legs meet the body (think: shoulders) and expel it out through large nozzles on their face. This creates the forceful front-facing current that the urine can be poured into. In addition, a fan organ near the mouth can blow this current sideways or backward. Pushing water backward pulls water in toward their face, allowing the lobster to smell what's ahead. Lobsters can also use their tail and smaller swimmerettes to blow water out behind them. The males often stand backward in the doorway of their dens and lift their tails as they shoot their urine scent out into the sea. Conversation with Jelle Atema, August 9, 2015; Jelle Atema and Molly A. Steinbach, "Chemical Communication and Social Behavior of the Lobster *Homarus americanus* and Other Decapod Crustacea," in *Evolutionary Ecology of Social and Sexual Systems: Crustaceans as Model Organisms,* ed. J. E. Duffy and M. Thiel (New York: Oxford University Press, 2007), 115–144.

Maine lobsters are serial monogamists: D. F. Cowan and J. Atema, 1990. "Moult staggering and serial monogamy in American lobsters, *Homarus americanus,*" *Anim. Behav.* 39 (1990): 1199–1206.

rapidly mature as a male first: D. Proestou, M. R. Goldsmith, and S. Twombly, "Patterns of male reproductive success in *Crepidula fornicata* provide new insight for sex allocation and optimal sex change," *Biol. Bull.* 214 (2008):194-202.

sometimes within less than two months: P. N. J. Chipperfield, "The breeding of *Crepidula fornicata* in the river Blackwater, Essex," *Journal of the Marine Biological Association of the UK* 30(1) (1951): 49–70.

impressively faithful to their partners: A. C. J. Vincent, and L. M. Sadler, "Faithful pair bonds in wild seahorses, *Hippocampus whitei,*" *Animal Behavior* 50 (1995): 1557–1569. Of course, other seahorse species have proven far less romantic, with males having broods mothered by different females in a single season. See Kvarnemo et al. "Monogamous pair bonds and mate switching in the Western Australian seahorse *Hippocampus subelongatus,*" *Journal of Evolutionary Biology* 13 (2000):882–888.

morning courtship ritual: C. M. C. Woods, "Preliminary observations on breeding and rearing the seahorse *Hippocampus abdominalis* (Teleostei: Syngnathidae) in captivity," *New Zealand Journal of Marine and Freshwater Research* 34(3) (2000): 475–485, doi:10.1080/00 288330.2000.9516950.

For the next several months: Vincent and Sadler, "Faithful pair bonds."

Dried seahorses are still sold: Woods, "Preliminary observations."

Consider the peacock wrasse: Dr. Bob Warner, professor of marine biology at University of California, Santa Barbara, proved a wealth of information with regard to the sexual strategies of wrasses and reef fish in general. His stories provide the foundation for several examples in this book, including the peacock wrasse, bluehead wrasse, and bucktooth parrotfish. Telephone interview by the author, November 19, 2014.

females give a guy a chance: Sarah B. M. Kraak and Eric P. Van Den Berghe, "Do female fish assess paternal quality by means of test eggs?" *Animal Behavior* 43 (1992): 865–867, and Andrea Manica, "Female scissortail sergeants (Pisces: Pomacentridae) use test eggs to choose good fathers," *Animal Behavior* 79(1) (2010): 237–242, doi:10.1016/j.anbehav.2009.11.006.

Large "pirate" males: E. P. van den Berghe, "Piracy as an alternative reproductive tactic for males," *Nature* 334(6184) (1988): 697–698.

Log enough hours: Warner has likely spent more time watching wrasses reproduce than any other person on the planet, perhaps more than even most wrasses.

no one builds as spectacular a nest: H. Kawase, Y. Okata, and K. Ito, "Role of huge geometric circular structures in the reproduction of a marine pufferfish," *Sci Rep* 3(2106) (2013), doi:10.1038/srep02106.

shaking his butt: The BBC has some marvelous footage of these pufferfish from their series *Life Story,* "Episode Five: Courtship." A clip is available online at http://www.bbc.co.uk/programmes/p029nb9g.

scientists confirmed: Keiichi Matsuura, "A new pufferfish of the genus *Torquigener* that builds 'mystery circles' on sandy bottoms in the Ryukyu Islands, Japan (Actinopterygii: Tetraodontiformes: Tetraodontidae)," *Ichthyol Res* 62 (2015): 207–212.

a healthy set of anal glands: M. Pizzolon et al., "When fathers make the difference: Efficacy of male sexually selected antimicrobial glands in enhancing fish hatching success," *Functional Ecology* 24 (2010): 141–148.

Dr. Phil Hastings: Hastings is curator of the Marine Vertebrate Collection at Scripps Institution of Oceanography, telephone interview by the author, February 25, 2015.

nearly all of them lay eggs: For a thorough review of blenny reproduction, see P. A. Hastings and C. W. Petersen, "Parental Care, Oviposition Sites, and Mating Systems in Blennioids," in *Reproduction and Sexuality in Marine Fishes: Patterns and Processes,* ed. K. S. Cole (Berkeley, University of California Press, 2010), 91–116.

a male redlip blenny takes stock: I. M. Côté and W. Hunte, "Self-monitoring of reproductive success: Nest switching in the redlip blenny (Pisces: Blenniidae)," *Behavioral Ecology and Sociobiology* 24 (1989): 403–408.

females won't compromise: P. Hastings, "Correlates of male reproductive success in the browncheek blenny, *Acanthemblemaria crockery* (Blennioidea: Chaenopsidae)," *Behav. Ecol. Sociobiol.* 22 (1988): 95–102.

Females are choosy: M. B. Rasotto, Y. Sadovy De Mitcheson, and G. Mitcheson, "Male body size predicts sperm number in the mandarinfish," *Journal of Zoology* 281(3) (2010): 161–167, doi:10.1111/j.1469-7998.2009.00688.x.

We like to ogle: Yvonne Sadovy, George Mitcheson, and Maria B. Rasotto, "Early development of the mandarinfish, *Synchiropus splendidus* (Callionymidae), with notes on its fishery and potential for culture," *Aquarium Sciences and Conservation* 3(4) (2001): 253-263.

Hastings notes: Conversation with the author, February 25, 2015.

the more sexually aroused a male gets: Licia Casaretto, Marta Picciulin, and Anthony D. Hawkins, "Mating behaviour by the haddock (*Melanogrammus aeglefinus*)," *Environmental Biology of Fishes* 98(3) (2104): 913–923, doi:10.1007/s10641-014-0327-7.

press their vents closely together: Like the vast majority of fish, haddock and cod are external fertilizers, releasing both eggs and sperm into the water column. In many species, pair-spawning is a common strategy for helping close the gap between sperm and egg in a swirling sea. More on how all that works in Act III.

Cod sex also culminates: Jeffrey A Hutchings, Todd D. Bishop, and Carolyn R. McGregor-Shaw, "Spawning behaviour of Atlantic cod, *Gadus morhua:* Evidence of mate competition and mate choice in a broadcast spawner," *Can. J. Fish Aquat Sci* 56 (1999): 97–104.

different fishing techniques can cramp: Sherrylynn Rowe and Jeffrey A. Hutchings, "Mating systems and the conservation of commercially exploited marine fish," *Trends in Ecology & Evolution* 18(11) (2003): 567–572, doi:10.1016/j.tree.2003.09.004.

all males in a population sing the same song: Daryl J. Boness, Phillip J. Clapham, and Sarah L. Mesnick, 2002. "Life History and Reproductive Strategies," in *Marine Mammal Biology: An Evolutionary Approach,* ed. A. Rus Hoelzel (Oxford: Blackwell Science, 2002), 278–324.

humpback whales also produce hit singles: Ellen C. Garland et al., "Dynamic horizontal cultural transmission of humpback whale song at the ocean basin scale," *Current Biology* (2011), doi:10.1016/j.cub.2011.03.019.

there are about four males to every female: K. C. Hall and R. Hanlon, "Principal features of the mating system of a large spawning aggregation of the giant Australian cuttlefish *Sepia apama* (Mollusca: Cephalopoda)," *Marine Biology* 140(3) (2002): 533–545, doi:10.1007/s00227-001-0718-0.

to easily fool a bigger male: R. T. Hanlon et al., "Transient sexual mimicry leads to fertilization," *Nature* 433 (2005): 212.

deception is taken a half step further: C. Brown, M. P. Garwood, and J. E. Williamson, "It pays to cheat: Tactical deception in a cephalopod social signaling system," *Biology Letters* 8 (2012): 729–732.

females fake having male parts: Elizabeth Preston summarizes this fascinating feature in her article "For disguise, female squid turn on fake testes," *Inkfish* (blog), *Discover,* October 10, 2014, http://blogs.discovermagazine.com/inkfish/2014/10/10/#.VdF2lixViko.

developing absolutely enormous testes: Robert R. Warner, "Synthesis: Environment, Mating Systems, and Life History Allocations in the Bluehead Wrasse," in *Model Systems in Behavioral Ecology: Integrating Conceptual, Theoretical, and Empirical Approaches,* ed. L. A. Dugatkin (Princeton, NJ: Princeton University Press, 2001), 227–244.

their testes shrink way down: This is an example of some of the colorful details provided by Warner during our conversation about bluehead wrasse, November 19, 2014.

there are simply too many wrasse: Warner, "Synthesis."

CHAPTER THREE: FLEX YOUR SEX

in some situations: Many of the details about how sex change works in marine fish came from my interviews of Dr. Robert Warner of University of California, Santa Barbara, and Dr. Matthew Grober of Georgia State University. For an overview of the theory of sex change, see Roldan C. Muñoz and Robert R. Warner, "A new version of the size-advantage hypothesis for sex change: Incorporating sperm competition and size-fecundity skew," *American Naturalist* 161(5) (2003): 749–761, and Robert R. Warner, "The adaptive significance of sequential hermaphroditism in animals," *American Naturalist* 109(965) (1975): 61–82.

he can continue to make babies: Recent studies, however, do show that male fertility declines with age too.

you would switch to being a male: Just to clarify, in sequential hermaphrodites, the switch from female to male is not because females are running dry on eggs—as we will discuss, fish and many invertebrates make more eggs as they grow. Instead, the switch happens because females gain more reproductive output as males for other reasons. Read on.

Fish are an exception: Dr. Matthew Grober has a developing hypothesis about why this is. Until it is published, I can't say too much at this point, but here's a hint: it may have to do with our kidneys. A similar idea was proposed by Homer Spencer in his book *From Fish to Philosopher* (New York: Little Brown, 1953).

for the opposite gender?: I took a little artistic license here in employing the word *gender.* Scientists studying the field of sex and gender stick to rather strict protocols in using *gender* only with reference to people.

in body and behavior, inside and out: Robert H. Devlin and Yoshitaka Nagahama, "Sex determination and sex differentiation in fish: An overview of genetic, physiological, and environmental influences," *Aquaculture* 208 (2002): 191–364.

Unsung defenders of coral reefs: The importance of parrotfish in preserving healthy coral reef ecosystems was highlighted in J. B. C. Jackson et al., eds., *Status and Trends of Caribbean*

Coral Reefs 1970–2012 (Gland, Switzerland: Global Coral Reef Monitoring Network, IUCN, 2012). This concept is a major component of new ocean management programs such as that by the Blue Halo Initiative discussed in chapter 8.

Sinewy, snakelike ribbon eels: L. Fishelson, 1990. "*Rhinomuraena spp.* (Pisces: Muraenidae): The first vertebrate genus with post-anally situated urogenital organs," *Marine Biology* 105(2) (1990): 253–257.

but also a crested cranium: The crest increases in size as does the size of the fish; since females switch to male at a total length of only about 30 to 35 inches, they don't sport this broadened brow. Dr. Yvonne Sadovy, in e-mail correspondence with the author, August 12, 2015.

a woman has all the eggs: This has been the traditional understanding of egg production in women for decades. However, recent studies hint that egg production may—to a limited extent—continue during adulthood. For more information, see the University of Michigan's Comprehensive Cancer Center website, http://www.mcancer.org/fertility-preservation/for -female-patients/normal-ovarian-function, and J. Goodwin, "Stem Cell Finding Could Expand Women's Lifetime Supply of Eggs," *Health Day,* February 26, 2012, http://consumer .healthday.com/senior-citizen-information-31/misc-aging-news-10/stem-cell-finding -could-expand-women-s-lifetime-supply-of-eggs-662135.html.

before the old gear is reabsorbed: Matthew Grober, in correspondence with the author, August 4, 2015.

a fourteen-inch-long vermilion snapper: Andrew Revkin relates a conversation with then-head of NOAA Jane Lubchenco in his article "Fishing Lessons," *New York Times,* September 1, 2009.

hedge their bets: M. A. Hixon, D. W. Johnson, and S. M. Sogard, "BOFFFFs: On the importance of conserving old-growth age structure in fishery populations," *ICES Journal of Marine Science* 71(8) (2014): 2171–2185.

hide within their waving tendrils: A mucus coating that covers the clownfish helps protect them from the anemone's sting.

suspended maturation: This suppression can also happen between two different species of clownfish co-occupying one anemone. See Akihisa Hattori, "Social and mating systems of the protandrous anemonefish *Amphiprion perideraion* under the influence of a larger congener," *Austral Ecology* 25 (2000): 187–192.

each individual knows his place in the pecking order: In one species of clownfish at least, subordinates reduce their growth rates in order to maintain enough size difference between themselves and the next-ranking fish. With a big enough size gap, the subordinate is not viewed as a threat, and conflict is likely reduced. P. Buston, "Size and growth modification in clownfish," *Nature* 424 (2003): 145–146.

can quickly convert to female: In some cases, it only takes about three weeks. Margarida Casadevall et al., "Histological study of the sex-change in the skunk clownfish *Amphiprion akallopisos,*" *Open Fish Science Journal* 2 (2009): 55–58.

Glued together as a living rock wall, oyster reefs or "beds": Some of the biggest reefs are those formed by the eastern (aka American) oyster along the eastern seaboard of the United States and the Gulf of Mexico, but not all species of oysters form large reefs.

During a season of summer lovin': Depending on location along the eastern seaboard of the United States, oysters may spawn from April through to October. Dr. Juliana Harding, Coastal Carolina University, in correspondence with the author, August 17, 2015.

with bigger oysters able: For details on sex change in oysters, I am grateful to insights shared via e-mail correspondence with Dr. Juliana Harding of Coastal Carolina University and Dr. Mark Luckenbach from the Virginia Institute of Marine Science.

seems to be mostly genetic: Ximing Guo, "genetic determinants of protandric sex in the Pacific oyster, *Crassostrea gigas* Thunberg," *Evolution* 52(2) (1998): 394–402.

may trigger an earlier transition to female: J. M. Harding et al., "Variation in eastern oyster (*Crassostrea virginica*) sex-ratios from three Virginia estuaries: protandry, growth, and demographics," *Journal of the Marine Biological Association of the United Kingdom* 93(Special Issue 2) (2012): 519–531.

owning killer real estate: R. Ross, "The evolution of sex-change mechanisms in fishes," *Environmental Biology of Fishes* 29 (1990): 81–93.

within an hour: Yasuhiro Nakashima et al., "Female-female spawning and sex change in a haremic coral-reef fish, *Labroides dimidiatus*," *Zoological Science* 17(7) (2000): 967–970, doi:10.2108/zsj.17.967.

sea goldie: Douglas Y. Shapiro, "Serial female sex changes after simultaneous removal of males from social groups of a coral reef fish," *Science* 209 (1980): 1136–1137.

takes the math skills a step further: R. C. Muñoz and R. R. Warner, "Testing a new version of the size-advantage hypothesis for sex change: Sperm competition and size-skew effects in the bucktooth parrotfish, *Sparisoma radians*," *Behav. Ecol.* 15 (2004): 129–136.

The female rusty angelfish is anything but rusty: Yoichi Sakai, "Alternative spawning tactics of female angelfish according to two different contexts of sex change," *Behavioral Ecology* 8(4) (1997): 372–377.

your favorite shrimp: A theoretical study of the popular food shrimp, *Pandalus,* noted that sex changing populations were more vulnerable to overfishing than non-sex-changing populations. In C. Fu, T. J. Quinn II, and T. C. Shirley, "The role of sex change, growth, and mortality in *Pandalus* population dynamics and management," *ICES Journal of Marine Science* 58 (2001): 607–621.

in a population and cause . . . shrinkage: For a great overview of human-driven shrinkage across the animal kingdom, see David Malakoff, "Shrink to Fit," *Conservation Magazine,* March 3, 2011.

has led to reductions: It is important to note here that not all of the detected changes in size and growth are genetically based—within any individual there is a certain degree of "plasticity," with traits such as growth rate or size at maturity displaying a normal range of values. Many of the responses to fishing thus far may be demonstrations of this individual plasticity; however, long-term selection from fishing may be enough to push species to evolve—meaning, have genetically based—reductions in growth rates and size at maturity that go beyond individual adaptations. For more details, see Marie-Joëlle Rochet, "Short-term effects of fishing on life history traits of fishes," *ICES Journal of Marine Science* 55 (1998): 371–391; M. R. Walsh et al., "Maladaptive changes in multiple traits caused by fishing: Impediments to population recovery," *Ecol Lett* 9(2) (2006): 142–148, doi:10.1111/j.1461-0248.2005.00858.x; D. O. Conover et al., "Reversal of evolutionary downsizing caused by selective harvest of large fish," *Proceedings of the Royal Society,* B. 276 (2009): 2015–2020.

happening with eastern oysters: J. M. Harding et al., "Variation in eastern oyster (*Crassostrea virginica*) sex-ratios." In addition to reduced reproductive output, this decline in large oysters may also restrict the available habitat for oyster larvae to settle upon, further threatening survival of the population.

Ophryotrocha puerilis: Anders Berglund, 1986. "Sex change by a polychaete: Effects of social and reproductive costs," *Ecology* 67(4) (1986): 837–845.

once you know the rules: E. W. Rodgers, R. L. Earley, and M. S. Grober, "Social status determines sexual phenotype in the bi-directional sex changing bluebanded goby *Lythrypnus dalli*," *Journal of Fish Biology* 70(6) (2007): 1660–1668, doi:10.1111/j.1095-8649.2007.01427.x.

But here is where bluebanded gobies get a bit crazy: I relied on a phone interview (April, 6, 2014) and follow-up correspondence with Matthew Grober for the details behind the process of this remarkable sex-changing fish.

It is now widely believed: Sibel Bargu et al., "Mystery behind Hitchcock's birds," *Nature Goescience* 5 (2013): 2–3. *The Birds* was also based on a short story by Daphne du Maurier.

eating too many diatoms: Valerio Zupo et al., "Do benthic and planktonic diatoms produce equivalent effects in crustaceans?," *Marine and Freshwater Behaviour and Physiology* 40(3) (2007): 169-181, doi:10.1080/10236240701592930.

Ford has found: The complex interaction of pollution and parasites in these coastal amphipods was explained in patient detail by Dr. Alex Ford through phone conversations (May 7, 2014) and e-mail exchanges with the author, and I also consulted his published works. For a great overview of parasites controlling their hosts' behaviors and bodies, check out Carl Zimmer's excellent *Parasite Rex* (New York: Free Press, 2000).

found two kinds of parasites: S. Short et al., "Paramyxean-microsporidian co-infection in amphipods: Is the consensus that Microsporidia can feminise their hosts presumptive?," *Int J Parasitol* 42(7) (2012): 683–691, doi:10.1016/j.ijpara.2012.04.014.

Previous studies have shown: For example, see L. L. Johnson et al., "Xenoestrogen exposure and effects in English sole (*Parophrys vetulus*) from Puget Sound, WA," *Aquat Toxicol* 88(1) (2008): 29–38, doi:10.1016/j.aquatox.2008.03.001, and Joseph G. Vos et al., "Health effects of endocrine-disrupting chemicals on wildlife," *Critical Reviews in Toxicology* 30(1) (2000): 71–133.

the most infamous example: D. Santillo, P. Johnson, and W. Langston, "Tributyltin (TBT) Antifoulants: A Tale of Ships, Snails and Imposex," in *The Precautionary Principle in the 20th Century: Late Lessons from Early Warnings,* eds. P. Harremoës et al. (New York: Earthscan, 2002), 148–160.

highest numbers of intersex amphipods: A. T. Ford et al., "Can industrial pollution cause intersexuality in the amphipod, Echinogammarus marinus?," *Mar Pollut Bull* 53(1–4) (2006): 100–106, doi:10.1016/j.marpolbul.2005.09.040.

CHAPTER FOUR: THE PENIS CHAPTER

***nearly* internal fertilization:** The technical term is "confined internal fertilization" as the sperm packets are deposited outside of the female's reproductive tract.

But it's a dangerous trek: John E. Randall, "Contributions to the biology of the queen conch, *Strombus gigas*," *Bulletin of Marine Science* 14(2) (1964): 246–295.

each one bent on inseminating the other: N. K. Michiels and L. J. Newman, "Sex and violence in hermaphrodites," *Nature* 391 (February 1998): 647.

sex with oneself isn't always a mild affair: S. A. Ramm et al., 2015. "Hypodermic self-insemination as a reproductive assurance strategy," *Proc Biol Sci* 282(1811) (2015), doi:10.1098/rspb.2015.0660.

enormous sex chains: Julie Ann Miller, "Sex and the sea hare," *Science News* 116 (1979): 218–219.

boosting success of sperm: R. Lange et al., "Functions, diversity, and evolution of traumatic mating," *Biol Rev Camb Philos Soc* 88(3) (2013): 585–601, doi:10.1111/brv.12018.

wiggle their way to the site of fertilization: Stabbing doesn't always work this way. For some species, insemination can only happen if the penis punctures a mate in a specific location on the body; in other species, it doesn't seem to matter as much. See Michiels and Newman, "Sex and violence in hermaphrodites," and L. Angeloni, "Sexual selection in a simultaneous hermaphrodite with hypodermic insemination: Body size, allocation to sexual roles, and paternity," *Animal Behavior* 66 (2003): 417–426.

as pairs of *Siphopteron sp.*: R. Lange, J. Werminghausen, and N. Anthes, "Cephalo-traumatic secretion transfer in a hermaphrodite sea slug," *Proc Biol Sci* 281(1774) (2014). doi:10.1098/rspb.2013.2424.

attempt to suck out: L. Schärer, G. Joss, and P. Sandner, "Mating behaviour of the marine turbellarian *Macrostomum sp.:* These worms suck." *Mar Biol* 145 (2004): 373–380.

It's far better, energetically, to be the inseminator: This is generally true but there are a few cases where the "female" side can benefit, such as if ejaculates contain nutrients, or can boost fertility; or when multiple mating opportunities arise, sperm competition or female choice may allow females to up their offspring fitness and thus, the female may encourage being stabbed by additional partners. R. Lange et al., "Female fitness optimum at intermediate mating rates under traumatic mating," *PLoS ONE* 7(8) (2012): e43234, doi:10.1371/journal.pone.0043234.

Such vacuuming up of seed: Ed Yong provides a wonderful synthesis of simple vs. weaponized sperm in his article "The sexual battles of flatworms: Barbed sperm, mating rings, traumatic insemination, and going down on yourself" on his blog *Not Exactly Rocket Science,*

January 10, 2011, http://blogs.discovermagazine.com/notrocketscience/2011/01/10/the
-sexual-battles-of-flatworms-barbed-sperm-mating-rings-traumatic-insemination-and
-going-down-on-yourself-2/#.VdpZRLxVikp, based on the paper by Lukas Schärer et
al., "Mating behavior and the evolution of sperm design," *PNAS* 108(4) (2011):1490–
1495.

like cheese graters: John Long, professor of palaeontology at Flinders University in Ad-
elaide, quoted in Ian Sample, "First act of sexual intercourse 'was done sideways, square-
dance style,'" *Guardian* (UK), October 14, 2014.

"astounding swimmer with a large penis": David J. Siveter et al., "An ostracode crustacean with soft parts from the Lower Silurian," *Science* 302 (2003): 1749–1751.

each species has its own pattern of sparks: James G. Morin and Anne C. Cohen, "It's all
about sex: Bioluminescent courtship displays, morphological variation and sexual selection
in two new genera of Caribbean ostracodes," *Journal of Crustacean Biology* 30(1) (2010):
56–67. doi:10.1651/09-3170.1.

into Van Gogh's *Starry Night*: After watching clips online of the bioluminescence of
ostracods, this seemed an appropriate attribute. Turns out, I'm not alone. Jennifer Fraz-
er's wonderful article, "The Starry Night Beneath the Caribbean Sea," provides beautiful description and images of courting ostracods via the *Artful Aomeba* blog on *Scientific
America*, October 20, 2014, http://blogs.scientificamerican.com/artful-amoeba/the-starry
-night-beneath-the-caribbean-sea/.

***Luprisca incuba*:** David J. Siveter et al., "Exceptionally preserved ostracods with develop-
mental brood care from the Ordovician," *Current Biology* 24(7) (2014): 801–806.

takes some serious engineering: For a great summary of the history and detailed physiology of penises across the animal kingdom, I highly recommend Barbara M. D. Natterson-
Horowitz and Kathryn Bowers's book, *Zoobiquity: What Animals Can Teach Us About Health
and the Science of Healing* (New York: Alfred A. Knopf, 2012).

Female mosquitofish find bigger gonopodia more attractive: R. B. Langerhans, C. A.
Layman, and T. J. DeWitt, "Male genital size reflects a tradeoff between attracting mates
and avoiding predators in two live-bearing fish species," *Proc Natl Acad Sci USA* 102(21)
(2005): 7618–7623, doi:10.1073/pnas.0500935102.

same basic engineering as a puffer: Natterson-Horowitz and Bowers, *Zoobioquity,* 63.

combined with a complex social system: For example, in eastern spinner dolphins, only
a fraction of males appear to participate in mating—remove these select individuals, and
it could have disproportionately large effects on reproductive output. Wade, Reeves, and
Mesnick, "Social and behavioural factors," 9.

such as sculpin: Not all species are so minimally endowed, with some having rather large
genital papilla and others having none at all. In these fish, sperm are expelled into the ocean.
The female, in a maneuver a bit like a frog's tongue catching a fly, everts a small structure
coated with mucus that snags the sperm and pulls them back up inside her. M. Muñoz,
"Reproduction in Scorpaeniformes," in *Reproduction and Sexuality in Marine Fishes,* ed.
Kathleen S. Cole (Berkeley: University of California Press, 2010).

"The prosciformed penis": C. Darwin, "A Monograph of the Sub-class Cirripedia—the
Balanidae, Part 1," in *The Works of Charles Darwin* vol. 11–13 (New York: NYU Press,
1988), 23.

remarkably adept penis shape-shifters: C. J. Neufeld and A. R. Palmer, "Precisely proportioned: Intertidal barnacles alter penis form to suit coastal wave action," *Proc Biol Sci*
275(1638) (2008): 1081–1087, doi:10.1098/rspb.2007.1760.

In contrast, some deep-sea squid don't have a hectocotylus: For much of the details
about squid copulation and especially sperm transfer, I turned to the expertise of Dr. Henk-
Jan Hoving of the Helmholtz Centre for Ocean Research Kiel.

known to strangle: For a fantastic discussion of the ins and outs (and cannibalistic threat) of
octopus sex, I highly recommend Katherine Harmon Courage's entertaining piece on BBC
Earth Strange and Beautiful, "How Male octopuses avoid being eaten by hungry females,"
2015, http://www.bbc.com/earth/story/20150223-mysteries-of-cannibal-octopus-sex.

interlocking beaks: K. S. Bolstad, 2006. "Sexual dimorphism in the beaks of *Moroteuthis ingens* Smith, 1881 (Cephalopoda: Oegopsida: Onychoteuthidae)," *New Zealand Journal of Zoology* 33(4) (2006): 317–327, doi:10.1080/03014223.2006.9518459.

a female giant hooked squid: A. I. Arkhipkin and V. V. Laptikhovsky, "Observation of penis elongation in *Onykia ingens:* implications for spermatophore transfer in deep-water squid," *Journal of Molluscan Studies* 76(3) (2010): 299–300, doi:10.1093/mollus/eyq019.

captured on film two squid midcoitus: H. J. T. Hoving and M. Vecchione, "Mating behavior of a deep-sea squid revealed by in situ videography and the study of archived specimens," *Biol. Bull.* 223 (2012): 263–267.

Or in some cases, male to male: H. J. T. Hoving, S. L. Bush, and B. H. Robison, "A shot in the dark: Same-sex sexual behaviour in a deep-sea squid," *Biol Lett* 8(2) (2012): 287–290, doi:10.1098/rsbl.2011.0680.

squid sperm packets have a mind of their own: H. J. T. Hoving and V. Laptikhovsky, "Getting under the skin: Autonomous implantation of squid spermatophores," *Biol. Bull.* 212 (2007): 177–197.

he amputates the precious appendage: V. Laptikhovsky and A. Salman, "On reproductive strategies of the epipelagic octopods of the superfamily Argonautoidea (Cephalopoda: Octopoda)," *Marine Biology* 142 (2003): 321–326, doi:10.1007/s00227-002-0959-6.

Known as epitoky: I turned to invertebrate expert Dr. Greg Rouse of Scripps Institution of Oceanography for more details on the process of epitoky and worm reproduction. I conducted phone interviews and several follow-up e-mail correspondences between June 2011 and 2015.

The regrowing process: Greg Rouse, communication with the author, August 17, 2015.

it *becomes* them: Although the epitoke is a sperm delivery device, really, as with the females, males are more like swimming testes—sacs packed with sperm. Greg Rouse, communication with the author, August 17, 2015.

disposable-regrowable-penises: A. Sekizawa et al., "Disposable penis and its replenishment in a simultaneous hermaphrodite," *Biol Lett* 9(2) (2013), doi:10.1098/rsbl.2012.1150.

But for male sharks: For details about shark ejaculation and copulation, I consulted with Dr. Dean Grubbs of Florida State University via phone interview, April 28, 2015, followed by several e-mail exchanges.

Nearly half of them live: C. F. Cotton and R. D. Grubbs, "Biology of deep-water Chondrichthyans: Introduction," *Deep Sea Research II* 115 (2015): 1–10.

Nurse shark sex: H. L. Pratt and J. C. Carrier, "A review of elasmobranch reproductive behavior with a case study on the nurse shark, *Ginglymostoma cirratum*," *Environmental Biology of Fishes* 60 (2001): 157–188.

males have specialized teeth: Pratt and Carrier, "A review of elasmobranch reproductive behavior," 160.

CHAPTER FIVE: INNER CHAMBERS

Dr. Sarah Mesnick: In conversation with the author, August 7, 2015.

to do with a longstanding assumption: M. Ah-King, A. B. Barron, and M. E. Herberstein, "Genital evolution: Why are females still understudied?," *PLoS Biol* 12(5) (2014): e1001851, doi:0.1371/journal.pbio.1001851.

Cryptic female choice: There are myriad techniques for deploying cryptic female choice. As this chapter aims to demonstrate, the more we look, the more we find. T.R. Birkhead provides an accessible and humorous discussion of the subject in his book *Promiscuity* (Harvard University Press, 2000), where in addition to providing examples of diverse mechanisms that may lead to cryptic female choice, he also discusses the challenge inherent in proving that such sexual selection exists.

study a whale vagina: Technically, Sarah Mesnick and Dara Orbach are studying cetacean vaginas—those of whales, dolphins, and porpoises—but "whale vagina" is a bit catchier. I'll use this phrase throughout to refer to all cetacean vagina research by this team.

in these species with small testes: There appears to be a trade-off in males between invest-ment in battle gear, such as tusks, and testes size. Winning battles means males don't have to deal with sperm competition, so they don't have to have large testes. Instead they invest in growing weapons. James P. Dines et al., "A trade-off between precopulatory and postcopu-latory trait investment in male cetaceans," *Society for the Study of Evolution* 69(6) (2015): 1560–1572.

highly structured social systems: For example, in toothed whales recoveries from overfish-ing have been much slower than in baleen whales with less complex social systems. Wade, Reeves, and Mesnick, "Social and behavioural factors." See also Elizabeth Pennisi, "Adult Killer Whales Need Their Mamas," *Science News,* September 13, 2012, http://news.science mag.org/2012/09/adult-killer-whales-need-their-mamas.

all major groups of vertebrates and many invertebrates: W. V. Holt and R. E. Lloyd, "Sperm storage in the vertebrate female reproductive tract: How does it work so well?" *The-riogenology* 73 (2010): 713–722.

females that lay eggs over protracted time periods: Tim Birkhead, *Promiscuity: An Evolu-tionary History of Sperm Competition* (Cambridge, MA: Harvard University Press, 2000), 70.

as little as fifteen minutes!: Bec Crew, "Our Wisdom of Birds (Or What Happens in Their Female Parts)" *Scitable,* September 13, 2012, http://www.nature.com/scitable/blog /scholarcast/our_wisdom_of_birds_or.

used by female hawksbill sea turtles: K. P. Phillips et al., "Reconstructing paternal geno-types to infer patterns of sperm storage and sexual selection in the hawksbill turtle," *Mol Ecol* 22(8) (2013): 2301–2312, doi:10.1111/mec.12235.

such was the case for a female brownbanded bamboo shark: M. A. Bernal et al., "Long-term sperm storage in the brownbanded bamboo shark *Chiloscyllium punctatum,*" *J Fish Biol* 86(3) (2015): 1171–1176, doi:10.1111/jfb.12606.

sire offspring from the Great Beyond: A. López-Sepulcre et al., "Beyond lifetime repro-ductive success: The posthumous reproductive dynamics of male Trinidadian guppies," *Proc Biol. Sci B* 280(1763) (2013), doi:10.1098/rspb.2013.1116.

likely because our high internal body temperature: T. R. Birkhead and A. P. Moller, "Sexual selection and the temporal separation of reproductive events: Sperm storage data form reptiles, birds and mammals," *Biological Journal of the Linnean Society* 50 (1993): 295–311.

a female can control how much: A. Pilastro et al., "Cryptic female preference for colorful males in guppies," *Evolution* 58(3) (2004): 665–669.

female green sea turtles: Julie Booth and James A. Peters, "Behavioural studies on the green turtle (*Chelonia mydas*) in the sea," *Animal Behaviour* 20 (1972): 808–812, and Dr. Peter Dutton, sea turtle expert with NOAA, conversation with the author, April 10, 2015.

turned to the same techniques: D. D. Chapman et al., "The behavioural and genetic mating system of the sand tiger shark, *Carcharias taurus,* an intrauterine cannibal," *Biology Letters* 9 (2013): doi:10.1098/rsbl.2013.0003.

is far more grisly: D. D. Chapman, phone interview by the author, May 16, 2014.

its behavior is revealed: For those wanting to see these embryo-cannibals in action, search "sand tiger embryo cannibalism video" and you'll be rewarded with footage of these em-bryos—swimming and hunting—from inside the womb.

lemon sharks have extremely high rates: K. A. Felheim, S. H. Gruber, and M. V. Ashley, "Multiple paternity of a lemon shark litter," *Copeia* 2001(3): 781–786.

In a separate study: D. D. Chapman et al., "Predominance of genetic monogamy by fe-males in a hammerhead shark, *Sphyrna tiburo:* Implications for shark conservation," *Molecu-lar Ecology* 13 (2014): 1965–1974.

And what happens is this: K. A. Paczolt and A. G. Jones, "Post-copulatory sexual selec-tion and sexual conflict in the evolution of male pregnancy," *Nature* 464(7287) (2010): 401–404, doi:10.1038/nature08861.

"We were out tagging whales": Dr. Phillip Clapham, phone conversation with the au-thor, February 9, 2015; and Bruce Mate et al., "Observations of a female North Atlantic right whale (*Ebalaena glacialis*) in simultaneous copulation with two males: Supporting evidence for sperm competition," *Aquatic Mammals* 31(2) (2005): 157–160, doi:10.1578/AM.31.2.2005.157.

also may invite sex: S. L. Mesnick, "Sexual alliances: Evidence and evolutionary implications." in *Feminism and Evolutionary Biology,* ed. Patricia Gowaty (New York: Springer, 1997), 207-260.

blood vessels start to entwine: Theodore W. Pietsch, "Dimorphism, parasitism and sex: Reproductive strategies among deepsea ceratioid anglerfishes," *Copeia* 1976(4): 781–793, doi:10.2307/1443462.

once-independent swimmer: Not all male anglerfish are parasitic or such extreme dwarves. Some species have free-swimming males that, while smaller than the females, retain their independence.

"So perfect is the union": Charles Tate Regan (1925), as quoted in Pietsch, "Dimorphism, parasitism and sex," 781.

Osedax, **a deep-sea worm:** For the play-by-play account of the *Osedax* saga and other deep-sea worm sex stories, I relied upon interviews of and several e-mail exchanges with Dr. Greg Rouse between 2011 and 2015, in addition to several published papers, including G. W. Rouse, S. K. Goffredi, and R. C. Vrijenhoek, "Osedax: Bone-Eating Marine Worms with Dwarf Males," *Science* 305(5684) (2004): 668–671, doi:10.1126/science.1098650, and G. W. Rouse et al., "Acquisition of dwarf male "harems" by recently settled females of *Osedax roseus* n. sp. (Siboglinidae; Annelida)," *Biological Bulletin* 214(1) (2008): 67–82.

100,000 times smaller: Rouse, e-mail correspondence with the author, August 17, 2015.

several sperm wiggle their way: Birkhead, *Promiscuity,* 186–187.

shark born in captivity: D. D. Chapman et al, "Virgin birth in a hammerhead shark," *Biology Letters* 3 (2007): 425–427, doi:10.1098/rsbl.2007.0189.

female blacktip shark: Chapman et al., "Virgin birth in a hammerhead shark."

novel application of a genetic technique: Andrew T. Fields et al., "Facultative parthenogenesis in a critically endangered wild vertebrate," *Current Biology* 25(11) (June 2015), R439–R447, doi:10.1016/j.cub.2015.04.018.

critically endangered smalltooth sawfish: See the International Union for the Conservation of Nature (IUCN) Red List of Threatened Species, October 5, 2015, http://www.iucnredlist.org/details/18175/0.

heading efforts to understand: Dr. Jim Gelsleichter, phone interview with the author, August 8, 2014.

CHAPTER SIX: OCEANIC ORGIES

so much as the tides: For basic biology and spawning of grunion I turned to the California Department of Fish and Wildlife's webpage and references therein. See https://www.wildlife.ca.gov/Fishing/Ocean/Grunion#28352307-california-grunion-facts.

four hundred people per square mile: Alaska is excluded from this calculation. NOAA, "State of the Coast Report," March 2013, http://stateofthecoast.noaa.gov/features/coastal-population-report.pdf.

There are initiatives: These include efforts to reduce armoring of beaches and grunion-friendly beach grooming protocols such as those provided by the Beach Ecology Coalition, http://grunion.pepperdine.edu/beachecologycoalition/grunion-grooming-protocol2013.pdf.

horseshoe crabs come to shore: There are four species of horseshoe crabs around the world. This section focuses on the Atlantic horseshoe crab *Limulus polyphemus.* For an overview about horseshoe crab biology and mating, see the website of the Ecological Research & Development Group, a nonprofit dedicated to conservation of horseshoe crabs, www.horseshoecrab.org.

same neurotoxin as puffers: J. Kanchanapongkul and P. Krittayapoositpot, "An epidemic of tetradotoxin poisoning following ingestion of the horseshoe crab *Carcinoscorpius rotundicauda,*" *Southeast Asian J Trop Med Public Health* 26(2) (1995): 364–367.

The operation looks like: For a summary of the horseshoe crab biomedical industry and its impacts, see Alexis C. Madrigal, "The Blood Harvest," *Atlantic,* February 26, 2014, http://www.theatlantic.com/technology/archive/2014/02/the-blood-harvest/284078/.

A recent study: R. L. Anderson, W. H. Watson III, and C. C. Chabot, "Sublethal behavioral and physiological effects of the biomedical bleeding process on the American horseshoe crab, *Limulus polyphemus,*" *Biological Bulletin* 225 (2014): 137–151.

are the choice pickings: Natalie Angier, "Tallying the toll on an elder of the sea," *New York Times,* June 10, 2008. Note that in some states, harvest of female horseshoe crabs is now banned.

alternative bait: Delaware's Sea Grant College program is one example of ongoing research to experiment with alternative bait. See http://www.deseagrant.org/research/saving -horseshoe-crab-designing-more-sustainable-bait-regional-eel-and-conch-fisheries.

rearing horseshoe crabs in the lab: For a review of horseshoe crab culture, see R. H. Carmichael, and E. Brush, "Three decades of horseshoe crab rearing: a review of conditions for captive growth and survival," *Reviews in Aquaculture* 4 (2012):32–43.

high-profile busts: For a play-by-play of a 2013 chase, see David Goodman, "It's Dark, but We See You: Release the Horseshoe Crabs," *New York Times,* May 29, 2013.

At least seventeen different species: William D. Heyman and Björn Kjerfve, "Characterizations of transient multi-species reef fish spawning aggregation at Gladden Spit, Belize," *Bulletin of Marine Science* 83(3) (2008): 531–551.

its own variation on the theme: Dr. Brad Erisman and Dr. Yvonne Sadovy shared firsthand accounts of witnessing spawning aggregations with me via a phone interview (Erisman on March 23, 2015) and e-mail exchanges (Sadovy), that supplement my descriptions based on the literature.

streak toward the surface: Females may also dash off into the distant blue on a more horizontal trajectory.

a telltale sign that bluefin are below: Fen Montaigne, "Still Waters: The Global Fish Crisis," *National Geographic,* April 2007.

was considered trash fish: Trevor Corson, *The Zen of Fish: The Story of Sushi, From Samurai to Supermarket* (New York: HarperCollins, 2007).

$1.76 million: Malcom Foster, "Bluefin tuna sells for incredible record $1.76 million at Tokyo fish auction," *Huffington Post,* January 5, 2013, http://www.huffingtonpost.com /2013/01/05/bluefin-tuna-sells-for-incredible-record-tokyo-fish-auction_n_2415722.html.

since Phoenician times: U. Ganzedo et al., "What drove tuna catches between 1525 and 1756 in southern Europe?," *ICES Journal of Marine Science* 66 (2009): 1595–1604.

had plummeted so far: Modern *almadraba* fishing catches in Spain are reported as approximately 1,370 metric tons, which is approximately 3 million pounds of bluefin. Luis Ambrosio and Pablo Xandril, "The Future of the Almadraba sector—Traditional Tuna Fishing Methods in the EU," European Union Directorate General for Internal Policies, 2015, http://www.europarl.europa.eu/studies.

these industrial fleets were catching: J. Fromentin et al., "The spectre of uncertainty in management of exploited fish stocks: The illustrative case of Atlantic bluefin tuna," *Marine Policy* 47 (2014): 8–14.

fair amount of uncertainty around: Fromentin, et al., "The spectre of uncertainty."

CHAPTER SEVEN: SYNCHRONIZED SEX

even at this unicellular level: Both an egg and a sperm are only one cell each.

sea urchins are the go-to: For a great summary of the history of science and sea urchins, along with videos of fertilizations of sea urchin eggs, check out the Exploratorium's website, http://www.exploratorium.edu/imaging-station/research/urchin/story_urchin1.php.

here is what we now know about sea urchin sex: Dr. Don Levitan of Florida State University graciously shared his insights regarding sea urchin reproduction via phone interview with the author, January 23, 2013.

euphoria-causing chemicals of marijuana: Aaron Rowe, "Chem lab: Sea urchin eggs plus marijuana equal amazing new drugs," *Wired Science,* November 4, 2007, http://www.wired.com/2007/11/chem-lab-hybrid.

good food gets sea urchins in the mood: K. Reuter and D. R. Levitan, "Influence of sperm and phytoplankton on spawning in the echinoid *Lytechinus variegatus,*" *Biol. Bull.* 219(2010): 198–206.

linking leafy greens and sex: K. Annabelle Smith, "When lettuce was a sacred sex symbol," *Smithsonian,* July 16, 2013, http://www.smithsonianmag.com/arts-culture/when-lettuce-was-a-sacred-sex-symbol-12271795/?no-ist.

both males and females are equally dependent on chance: Don R. Levitan, "Density-dependent sexual selection in external fertilizers: Variances in male and female fertilization success along the continuum from sperm limitation to sexual conflict in the sea urchin *Strongylocentrotus franciscanus,*" *American Naturalist* 164(3) (2004): 298–309.

using these very old but still-kicking: D. R. Levitan, "Contemporary evolution of sea urchin gamete-recognition proteins: Experimental evidence of density-dependent gamete performance predicts shifts in allele frequencies over time," *Evolution* 66 (2012): 1722–1736.

Abalone off the West Coast of the United States: I turned to abalone expert Dr. Kristin Gruenthal for insight into reproduction and conservation management of this species during a phone interview, March 20, 2013.

Allee effects: Joanna Gascoigne et al., "Dangerously few liaisons: A review of mate-finding Allee effects." *Population Ecology* 51(3) (209): 355–372, doi:10.1007/s10144-009-0146-4.

we are raging sex addicts: Human sexual activity frequency from Kinsey Institute, http://www.kinseyinstitute.org/resources/FAQ.html.

they spawn at slightly different times: N. Fogarty et al., "Asymmetric conspecific sperm precedence in relation to spawning times in the *Montastraea annularis* species complex (Cnidaria: Scleractinia)," *Journal of Evolutionary Biology* 25 (2012): 2481–2488.

it's the gusto of the frank sperm: Dr. Fogarty provided additional details regarding the process of coral spawning and fertilization in an interview on February 5, 2013, and follow-up e-mails.

genetic factors and density: Don R. Levitan et al., "Genetic, spatial and temporal components of precise spawning synchrony in reef building corals of the *Montastraea annularis* species complex," *Evolution* 65(5) (2011): 1254–1270.

$375 billion annually: R. Costanza et al., "The value of the world's ecosystem services and natural capital," *Nature* 387 (1997): 253–260.

ACT III: POST-CLIMAX

the more gelatinous, slimy, and microbial in nature: Dr. Jeremy B. C. Jackson, my PhD adviser, has written extensively about what he calls "the rise of slime," whereby human interactions with the oceans are creating an environment more like the ancient primordial sea: great for life forms such as jellyfish, not so great for the species we tend to rely upon and favor most.

and find accommodation: In a nod to Rachel Carson, who writes in the closing chapter of *Silent Spring:* "Only by taking account of such life forces and by cautiously seeking to guide them into channels favorable to ourselves can we hope to achieve a reasonable accommodation between the insect hordes and ourselves."

CHAPTER EIGHT: TURNING UP THE SEX DRIVE

The oceans are far emptier: V. Christensen et al., "A century of fish biomass decline in the ocean," *Marine Ecology Progress Series* 512 (2014): 155–166, and J. B. C. Jackson et al., eds.,

Status and Trends of Caribbean Coral Reefs 1970–2012 (Gland, Switzerland: Global Coral Reef Monitoring Network, IUCN, 2012).

far more full of plastic particles and dead zones: Known as "dead zones," the number of areas of the ocean now carrying too little oxygen to support marine life exceeds four hundred, up from only four such zones in the 1900s. These zones are often caused by high levels of fertilizer runoff from lawns and farms that wash into the sea and fuel algal blooms. When the algae die, bacteria decomposing the algae suck up all the oxygen from the water. Since the 1960s, the number of dead zones has doubled every ten years. Climate change exacerbates the problem and is expected to increase the severity, duration, size, and impact of dead zones in coming years. R. Diaz and R. Rutger, "Spreading dead zones and consequences for marine ecosystems," *Science* 321(5891) (2008): 926–929.

#oceanoptimism: This was launched at World's Ocean Day on June 8, 2014, and stemmed from a think tank initiated by Dr. Nancy Knowlton to help the marine conservation community move, as she says, "beyond the obituaries" of nature.

has proven effective: H. M. J. Overzee and A. D. Rijnsdorp, "Effects of fishing during the spawning period: Implications for sustainable management," *Reviews in Fish Biology and Fisheries* 25 (2015): 65–83.

especially hard to "see" declines: Dr. Yvonne Sadovy of the University of Hong Kong calls this "the illusion of plenty." Correspondence with the author, August 14, 2015.

The case of red hinds: Richard S. Nemeth, "Population characteristics of a recovering US Virgin Islands red hind spawning aggregation following protection," *Marine Ecology-Progress Series* 286 (2005): 81–97.

a grassroots campaign called 4FJ: Sadovy noted the growing success of this campaign as an example of bottom-up efforts to protect spawning aggregations. Correspondence with the author, August 14, 2015.

invasive species that need culling: We haven't discussed invasive species, but they are a major problem for the health of the oceans. In contrast to the scenarios focused on here, with invasive species the problem is too much sex, rather than too little. These species flourish in their new environs, threatening local species through predation or competition for prey. This is a problem not just in the sea, but on land too, and it's generated interest from folks who turn to forks as a solution to the problem. Check out the recipe-heavy website led by conservation biologist Joe Roman, Eat the Invaders: fighting invasive species one bite at a time; see http://eattheinvaders.org.

some consider "ugly fish": Writer Clare Leschin-Hoar sums up the problem nicely in her recent article, "There's a price to pay for not eating America's ugly seafood," *Take Part,* June 29, 2015, http://www.takepart.com/article/2015/06/29/american-seafood-exports.

two-thirds of once-overfished populations: B. Sewell, 2013. "Bringing back the fish: An evaluation of U.S. fisheries rebuilding under the Magnuson-Stevens Fishery Conservation and Management Act," National Resources Defense Council, 2013, http://www.nrdc.org/oceans/rebuilding-fisheries.asp.

Diversifying which species we demand: The idea of diversifying seafood consumption is catching on. Paul Greenberg, in his *New York Times* article "Three Simple Rules for Eating Seafood," June 13, 2015, gave a nod to Michael Pollan's rules for sustainable food consumption and wrote: "Eat American Seafood; A much greater variety than we currently do; Mostly farmed filter feeders." Another organization, Chefs Collaborative, has held annual "Trash Fish" dinners for years to encourage such diversified palettes; see http://www.chefs collaborative.org/programs/seafood-solutions/trash-fish-dinners/.

one of you: K. Warner et al., "Oceana study reveals seafood fraud nationwide," *Oceana,* February 2013, http://oceana.org/reports/oceana-study-reveals-seafood-fraud-nationwide.

caught by slaves at sea: Recent exposés by the *Guardian* (2013) and the *New York Times* (2015) have uncovered the hidden world of slavery and abuse that occurs on many offshore fishing fleets. The catch from these vessels is often turned into feed for aquaculture, livestock, or pet food products—the first two categories fueling fish and meat we then eat. See K. Hodal, C. Kelly, and F. Lawrence, "Revealed: Asian slave labour producing prawns for supermarkets in US, UK," *Guardian* (UK), June 10, 2014, and I. Urbina, "'Sea Slaves': The human misery that feeds pets and livestock," *New York Times,* July 27, 2015.

Storied Fish: The idea of "storied fish" is part of the foundation of the work I do as research co-director at Future of Fish. The concept was first conceived by Future of Fish founder and CEO of Flip Labs LLC, Cheryl Dahle, as a key to improving sustainability of the global seafood supply chain during an initial research project that started in 2008. The theory of change—to improve traceability and bring more storied fish to market—remains the core focus of our work today. Find out more at www.futureoffish.org.

Between $10 and $23 billion worth: Fact sheet, "Illegal Fishing: Not in Our Ports," Office of International Affairs, NOAA, http://www.nmfs.noaa.gov/ia/iuu/portstate_factsheet .pdf, accessed on September 29, 2015.

Community Supported Fisheries (CSFs): For more information about locally sourced fish and to find CSFs near you, check out www.localcatch.org.

CSFs have also been shown: L. McClenachan et al., "Do community supported fisheries (CSFs) improve sustainability?," *Fisheries Research* 157 (2014): 62–69.

3-D ocean farming: You can find out more about how 3-D ocean farming works at Thimble Island Oyster's website, http://www.thimbleislandoysters.com/1379-2/.

an estimated *five pounds*: NOAA Fisheries Feeds for Aquaculture FAQ, http://www.nmfs .noaa.gov/aquaculture/faqs/faq_feeds.html#17how, accessed Jun 30, 2015.

pressure to turn these fish into feed: For a summary of the issue see Tom Philpott, "Eat a sardine, save a salmon," *Mother Jones,* November 9, 2011.

scientists caught a break: I received recent updates on the state of white abalone recovery and rearing from Dr. Kristin Aquilino in a phone interview, September 16, 2015.

nearly the size of Germany: Great Barrier Reef Marine Park Authority, "About the Reef," http://www.gbrmpa.gov.au/about-the-reef/facts-about-the-great-barrier-reef, accessed June 30, 2015.

The first time scientists observed: R. C. Muñoz et al., "Extraordinary aggressive behavior from the giant coral reef fish, *Bolbometopon muricatum,* in a remote marine reserve," *PLoS One* 7(6) (2012): e38120, doi:10.1371/journal.pone.0038120.

accessible "spyware": For background on the development of satellite-based data for enforcement and management, I relied on the expertise of SkyTruth founder John Amos in a telephone conversation, February 13, 2015.

Approximately 63 million sharks: Boris Worm et al., "Global catches, exploitation rates, and rebuilding options for sharks," *Marine Policy* 40 (2013): 194–204, doi:10.1016/j. marpol.2012.12.034.

shark fins are dried and shipped: A. T. Fields et al., "A novel mini-DNA barcoding assay to identify processed fins from internationally protected shark species," *PLoS ONE* 10(2) (2015): e0114844.

"mini–DNA barcoding" test: Fields et al., "A novel mini-DNA."

proved this one whale: Juan Pablo Torres-Florez et al., "First documented migratory destination for eastern South Pacific blue whales," *Marine Mammal Science* (August 2015): doi:10.1111/mms.12239.

the longest distance: Specifically, this was the longest distance over a latitudinal (north-south) gradient ever recorded.

Currently numbering around 490: This estimate is provided by the International Whaling Commission, "Whale Population Estimates," https://iwc.int/estimate#table, accessed September 2015.

a series of "smart" buoys that can detect whale calls: Cornell Lab of Ornithology Right Whale Listening Network, Bioacoustics Research Program, maintains an active website detailing this research, especially under the "Threats to Right Whales" section, http://www .listenforwhales.org/page.aspx?pid=439, accessed August 7, 2015.

These data show when the fish start to travel: L. Whaylen et al., "Observations of a Nassau grouper, *Epinephelus striatus,* spawning aggregation site in Little Cayman, Cayman Islands, including multi-species spawning information," *Environmental Biology of Fishes* 70 (2004): 305–313.

approximately double the number of fish since 2003: Brice Semmens, personal communication with the author, June 25, 2015.

ocean zoning: For a brief overview of ocean zoning, see the Waitt Institute's ocean zoning fact sheet (as well as fact sheets on other ocean topics) under the "Fact Sheets" page on their website: http://waittinstitute.org/factsheets/.

gained an astounding 463 percent more fish: O. Aburto-Oropeza et al., "Large recovery of fish biomass in a no-take marine reserve," *PLoS One* 6(8) (2011): e23601, doi:10.1371/journal.pone.0023601.

fewer than 5 percent of the individuals were males: Y. Sadovy de Mitcheson and B. Erisman, "Fishery and Biological Implications of Fishing Spawning Aggregations and the Social and Economic Importance of Aggregating Fishes," in *Reef Fish Spawning Aggregations: Biology, Research and Management, Fish & Fisheries Series 35,* eds. Y. Sadovy de Mitcheson and P. L. Colin (Netherlands: Springer, 2011), 225–284, doi: 10.1007/978-94-007-1980-4_8.

Female gag grouper ripe with eggs: Felicia C. Coleman, Christopher C. Koenig, and L. Alan Collins, "Reproductive styles of shallow-water groupers (Pisces: Serranidae) in the eastern Gulf of Mexico and the consequences of fishing spawning aggregations," *Environmental Biology of Fishes* 47 (1996): 129–141.

One potential approach: M. W. Russell, B. E. Luckhurst, and K. C. Lindeman, "Management of Spawning Aggregations," in *Reef Fish Spawning Aggregations: Biology, Research and Management,* Fish & Fisheries Series 35, eds. Y. Sadovy de Mitcheson and P. L. Colin (Netherlands: Springer, 2011), 371–401.

other forms of compensation and livelihood: Alternative incomes can also backfire, such as when the new income provides a stable resource that enables fishers to go out and fish more. This scenario happened in the Pacific Island of Kiribati and serves as a warning to ensure that whatever alternatives are provided, they are accompanied by appropriate regulations and monitoring to ensure reduced pressure is realized. For more on this, see Richard Harris, "Reef Conservation Strategy Backfires," National Public Radio, November 18, 2009, http://www.npr.org/templates/story/story.php?storyId=120536304.

a recent economic evaluation: G. M. S. Vianna et al., "Wanted dead or alive? The relative value of reef sharks as a fishery and an ecotourism asset in Palau," Australian Institute of Marine Science and University of Western Australia, Perth, 2010.

increased levels of endocrine-disrupting chemicals: L. L. Johnson et al., "Xenoestrogen exposure and effects in English sole (*Paraophrys vetulus*) from Puget Sounds, WA," *Aquatic Toxicology* 88 (2008): 29–38, and Joseph G. Vos et al., "Health effects of endocrine-disrupting chemicals on wildlife," *Critical Reviews in Toxicology* 30(1) (2000): ProQuest SciTech Collection, p. 71.

failure in harbor porpoises: S. Murphy et al., "Reproductive failure in UK harbour porpoises *Phocoena phocoena*: Legacy of pollutant exposure?," *PLoS ONE* 10(7) (2015): e0131085, doi:10.1371/journal.pone.0131085.

hormones used in the cattle industry: M. G. Bertrama, et al., "Sex in troubled waters: Widespread agricultural contaminant disrupts reproductive behavior in fish," *Hormone Behavior* 70 (2015): 85–91, doi: 10.1016/j.yhbeh.2015.03.002.

between eleven billion and twenty-six billion pounds: Jenna R. Jambeck et al., "Plastic waste inputs from land into the ocean," *Science* 347(6223) (February 13, 2015): 768–771.

more than 80 percent of plastic pollution: Jambeck et al., "Plastic waste inputs."

with penis growing on a downward trend: Santillo et al., "Tributyltin (TBT) Antifoulants."

in recent experiments on reef fish: J. M. Donelson and P. L. Munday, "Transgenerational plasticity mitigates the impact of global warming to offspring sex ratios," *Global Change Biology* 21(8) (2015): 2954–2962, doi:10.1111/gcb.12912.

abalone populations inside fully protected reserves: F. Micheli et al., "Evidence that marine reserves enhance resilience to climatic impacts," *PLoS ONE* 7(7) (2012): e40832, doi:10.1371/journal.pone.0040832.

show resistance to: Jackson et al., *Status and Trends of Caribbean Coral Reefs.*

West Coast rockfish: Russ Parsons, "Seafood Watch cites dramatic turnaround in rockfish, other West Coast fish," *Los Angeles Times,* September 2, 2014.

BIBLIOGRAPHY

The Bibliography contains a listing of additional works used by the author but not cited in the text or Notes section.

Akesson, Bertil, and John D. Costlow. "Effect of constant and cyclic temperatures at different salinity levels on survival and reproduction in *Dinophilus gyrociliatus* (Polychaeta: Dinophilidae)." *Bulletin of Marine Science* 48(2) (1991): 485–499.

Allsop, D. J., and S. A. West. "Constant relative age and size at sex change for sequentially hermaphroditic fish." *Journal of Evolutionary Biology* 16 (2003): 921–929.

Alonzo, S. H. "Uncertainty in territory quality affects the benefits of usurpation in a Mediterranean wrasse." *Behavioral Ecology* 15(2) (2004): 278–285, doi:10.1093/beheco/arh007.

Altieri, A. H., and K. B. Gedan. "Climate change and dead zones." *Global Change Biology* 21(4) (2015): 1395–1406, doi:10.1111/gcb.12754.

Aneloni, Lisa. "Sexual selection in a simultaneous hermaphrodite with a hypodermic insemination: Body size, allocation to sexual roles and paternity." *Animal Behavior* 66(3) (2003): 417–426, doi:10.1006/anbe.2003.2255.

Atkinson, S. "Male Reproductive Systems." In *Encyclopedia of Marine Mammals,* edited by W. F. Perrin, B. Wursig, and J. G. M. Thewissen. Waltham, MA: Academic Press, 2009.

Bagøien, Espen, and Thomas Kiørboe. "Blind dating—mate finding in planktonic copepods. I. Tracking the pheromone trail of Centropages typicus." *Marine Ecology Progress Series* 300 (2005): 105–115.

Baird, A. H., and P. A. Marshall. "Mortality, growth and reproduction in scleractinian corals following bleaching on the Great Barrier Reef." *Marine Ecology Progress Series* 237 (2002): 133–141.

Batchelor, S. N., D. Carr, C. E. Coleman, L. Fairclough, and A. Jarvis. "The photofading mechanism of commercial reactive dyes on cotton." *Dyes and Pigments* 59(3) (2003): 269–275, doi:10.1016/s0143-7208(03)00118-9.

Bauer, Raymond R. "Sex change and life history pattern in the shrimp *Thor Manningi* (decapoda: caridea): A novel case of partial protandric hermaphroditism." *Biol. Bull.* 170(1) (1986): 11–31.

Bergstrom, B. I. "Do protandric pandalid shrimp have environmental sex determination?" *Marine Biology* 128 (1997): 397–407.

Berkeley, Steven A., Colin Chapman, and Susan M. Sogard. "Maternal age as a determinant of larval growth and survival in a marine fish, *Sebastes melanops.*" *Ecological Society of America* 85(5) (2004): 1258–1264.

Bland, K. P., and C. C. Kitchener. "The anatomy of the penis of a sperm whale (*Physeter catodon* L., 1758)." *Mammal Rev.* 31(3) (2001): 239–244.

Boomer, J. J., R. G. Harcourt, M. P. Francis, T. I. Walker, J. M. Braccini, and A. J. Stow. "Frequency of multiple paternity in gummy shark, *Mustelus antarcticus,* and Rig, *Mustelus lenticulatus,* and the implications of mate encounter rate, postcopulatory influences, and reproductive mode." *J Hered* 104(3) (2013): 371–379, doi:10.1093/jhered/est010.

Bolstad, K. S. "Sexual dimorphism in the beaks of *Moroteuthis ingens* Smith, 1881(Cephalopoda: Oegopsida: Onychoteuthidae)." *New Zealand Journal of Zoology* 33(4) (2006): 317–327.

Borja, Ángel, María Jesús Belzunce, Joxe Mikel Garmendia, José Germán Rodríguez, Oihana Solaun, and Izaskun Zorita. "Impact of Pollutants on Coastal and Benthic Marine Communities." *Ecological Impacts of Toxic Chemicals* (2011): 165–186.

Boxshall, G. A. "Preface to the themed discussion on 'Mating biology of copepod crustaceans.'" *Royal Society* 353 (1998): 669–670, doi:10.1098/rstb.1998.0232.

Brady, A. K., J. D. Hilton, and P. D. Vize. "Coral spawn timing is a direct response to solar light cycles and is not an entrained circadian response." *Coral Reefs* 28(3) (2009): 677–680, doi:10.1007/s00338-009-0498-4.

Branch, G.M., and F. Odendaal. "The effects of marine protected areas on the population dynamics of a South African limpet, *Cymbula oculus*, relative to the influence of wave action." *Biol. Conserv.* 114 (2003): 255–269.

Bshary, Redouan, and Manuela Würth. "Cleaner fish *Labroides dimidiatus* manipulate client reef fish by providing tactile stimulation." *Royal Society* 268 (2001): 1495–1501.

Bulhneim, H. P. "Microsporidian infections of amphipods with special reference to host-parasite relationships: A review." *Marine Fisheries Review* 37(5–6) (1975): 39–45.

Bulseco, Ashley. "A synopsis of the Olympia oyster *Ostrea lurida*." *Aquaculture* 262 (1982): 63–72.

Buresch, K. C., M. R. Maxwell, M. R. Cox, and R. T. Hanlon. "Temporal dynamics of mating and paternity in the squid *Loligo pealeii*." *Marine Ecology Progress Series* 387 (2009): 197–203, doi:10.3354/meps08052.

Burton, R. S. "Mating system of the intertidal copepod *Tigriopus californicus*." *Marine Biology* 86(3) (1985): 247–252, doi:10.1007/BF00397511.

Bush, Stephanie L., Hendrik J. T. Hoving, Christine L. Huffard, Bruce H. Robison, and Louis D. Zeidberg. "Brooding and sperm storage by the deep-sea squid *Bathyteuthis berryi* (Cephalopoda: Decapodiformes)." *Journal of the Marine Biological Association of the United Kingdom* 92(07) (2012): 1629–1636, doi:10.1017/s0025315411002165.

Buston, Peter. "Forcible eviction and prevention of recruitment in the clown anemonefish." *Behavioral Ecology* 14(4) (2003): 576–582.

Byrne, Rosemary J., and John C. Avise. "Genetic mating system of the brown smoothhound shark (*Mustelus henlei*), including a literature review of multiple paternity in other elasmobranch species." *Marine Biology* 159(4) (2011): 749–756, doi:10.1007/s00227-011-1851-z.

Carpenter, Kent E., Muhammad Abrar, Greta Aeby, Richard B. Aronson, Stuart Banks, Andrew Bruckner, Angel Chiriboga, et al. "One-third of reef-building corals face elevated extinction risk from climate change and local impacts." *Science* 321 (June 2008).

Carrier, David R., Stephen M. Deban, and Jason Otterstrom. "The face that sank the *Essex:* Potential function of the spermaceti organ in aggression." *Journal of Experimental Biology* 205 (2002): 1755–1763.

Carrier, Jeffrey C., Harold L. Pratt Jr., and Linda K. Martin. "Group reproductive behaviors in free-living nurse sharks, *Ginglymostoma cirratum*." *American Society of Ichthyologists and Herpetologists* 3 (1994): 646–656.

Carrier, Jeffrey C., John A. Musick, and Michael R. Heithaus. Biology of Sharks and Their Relatives. 2nd, ed. Boca Raton, FL: CRC Press, 2012.

Carson, Rachel. *Silent Spring: With an Introduction by Vice President Al Gore.* New York: Houghton Mifflin, 1962.

Ceballos, S., and T. Kiorboe. "Senescence and sexual selection in a pelagic copepod." *PLoS One* 6(4) (2011): e18870, doi:10.1371/journal.pone.0018870.

Chabot, Chris L., and Brent M. Haggin. "Frequency of multiple paternity varies between two populations of brown smoothhound shark, *Mustelus henlei*." *Marine Biology* 161 (2014): 797–804.

Chapman, D. D., B. Firchau, and M. S. Shivji. "Parthenogenesis in a large-bodied requiem shark, the blacktip *Carcharhinus limbatus*." *Journal of Fish Biology* 73 (2008): 1473–1477, doi:10.1111/j.1095-8649.2008.02018.x.

Chapman, Demian D., Sabine P. Wintner, Debra L. Abercrombie, Jimiane Ashe, Andrea M. Bernard, Mahmood S. Shivji, and Kevin A. Feldheim. "The behavioural and genetic mating system of the sand tiger shark, *Carcharias taurus*, an intrauterine cannibal." *Biology Letters* 9 (2013), doi:10.1098/rsbl.2013.0003.

Charlton, B. D., Roland Frey, Allan J. McKinnon, Guido Fritsch, W. Tecumseh Fitch, and David Reby. "Koalas use a novel vocal organ to produce unusually low-pitched mating calls." *Current Biology* 23(23) (2013): R1035-R1036.

Cheung, C. H., J. M. Chiu, and R. S. Wu. "Hypoxia turns genotypic female medaka fish into phenotypic males." *Ecotoxicology* 23(7) (2014): 1260–1269, doi:10.1007/s10646-014-1269-8.

Chiba, S., K. Yoshino, M. Kanaiwa, T. Kawajiri, and S. Goshima. "Maladaptive sex ratio adjustment by a sex-changing shrimp in selective-fishing environments." *J Anim Ecol* 82(3) (2013): 632–41, doi:10.1111/1365-2656.12006.

Chipperfield, P. N. J. "The breeding of Crepidula Fornicata (L.) in the River Blackwater, Essex." *Journal of the Marine Biological Association of the United Kingdom* 30(1) (1951): 49–71.

Cigliano, John A. "Assessment of the mating history of female pygmy octopuses and a possible sperm competition mechanism." *Animal Behaviour* 49 (1995): 849–851.

Clarke, Tayler M., Mario Espinoza, and Ingo S. Wehrtmann. "Reproductive ecology of demersal elasmobranchs from a data-deficient fishery, Pacific of Coast Rica, Central America." *Fisheries Research* 157 (2014): 96–105.

Cole, Kathleen S. *Reproduction and Sexuality in Marine Fishes: Patterns and Processes.* Berkeley: University of California Press, 2010.

Colin, Patrick L. "Reproduction of the Nassau grouper, *Epinephelus striatus* (Pisces: Serranidae) and its relationship to environmental conditions." *Environmental Biology of Fishes* 34 (1992): 357–377.

Conover, D. O., and S. B. Munch. "Sustaining fisheries yields over evolutionary time scales." *Science* 297(5578) (2002): 94–96, doi:10.1126/science.1074085.

Corson, Trevor. *The Zen of Fish: The Story of Sushi, From Samurai to Supermarket.* New York: HarperCollins, 2007.

Corson, Trevor. *The Secret Life of Lobsters: How Fishermen and Scientist Are Unraveling the Mysteries of Our Favorite Crustacean.* New York: Harper Collins, 2004.

Côté, I. M., and W. Hunte. "Male and female mate choice in the redlip blenny: Why bigger is better." *Animal Behaviour* 38 (1989): 78–88.

Courage, Katherine Harmon. *Octopus: The Most Mysterious Creature in the Sea.* New York: Penguin Random House, 2013.

Crozier, L. G., and J. A. Hutchings. "Plastic and evolutionary responses to climate change in fish." *Evol Appl* 7(1) (2014): 68–87, doi:10.1111/eva.12135.

Cummins, J. M., and P. F. Woodall. "On mammalian sperm dimensions." *Journal of Reproduction & Fertility Ltd* 75 (1985): 153–175.

Dall, Sasha R. X., and Nina Wedell. "Evolutionary conflict: Sperm wars, phantom inseminations." *Current Biology* 15(19) (2005): R801–R803, doi:10.1016/j.cub.2005.09.019.

Daly-Engel, Toby S., R. Dean Grubbs, Kevin A. Feldeim, Brian W. Bowen, and Robert J. Toonen. "Is multiple mating beneficial or unavoidable? Low multiple paternity and genetic diversity in the shortspine spurdog *Squalus mitsukurii*." *Marine Ecology Progress Series* 403 (2010): 255–267.

Dam, Joost W. van, Andrew P. Negri, Sven Uthicke, and Jochen F. Mueller. "Chemical pollution on coral reefs: Exposure and ecological effects." In *Ecological Impacts of Toxic Chemicals.* Ed. F. Sánchez-Bayo, P. J. van den Brink and R. M. Mann. Oak Park, IL: Bentham Science Publishers, 2011: 187–211. doi: 10.2174/97816080512121110101.

DeMartini D. G., A. Ghoshal, E. Pandolfi, A. T. Weaver, M. Baum, and D. E. Morse. "Dynamic biophotonics: Female squid exhibit sexually dimorphic tunable leucophores and iridocytes." *Journal of Experimental Biology* 216(pt. 19) (2013): 3733–3741.

Devlin, Robert H., and Yoshitaka Nagahama. "Sex determination and sex differentiation in fish: An overview of genetic, physiological, and environmental influences." *Aquaculture* 208 (2002): 191–364.

Di Poi, C., A. S. Darmaillacq, L. Dickel, M. Boulouard, and C. Bellanger. "Effects of perinatal exposure to waterborne fluoxetine on memory processing in the cuttlefish *Sepia officinalis*." *Aquat Toxicol* 132–133(02) (2013): 84–91. doi:10.1016/j.aquatox.2013.02.004.

Dines, J. P., E. Otarola-Castillo, P. Ralph, J. Alas, T. Daley, A. D. Smith, and M. D. Dean. "Sexual selection targets cetacean pelvic bones." *Evolution* 68(11) (2014): 3296–3306, doi:10.1111/evo.12516.

Dines, James P., Sarah L. Mesnick, Katherine Ralls, Laura May-Collado, Ingi Agnarsson, and Matthew D. Dean. "A trade-off between precopulatory and postcopulatory trait investment in male cetaceans." *Society for the Study of Evolution* 69(6) (2015): 1560–1572.

Dixson, D. L., A. R. Jennings, J. Atema, and P. L. Munday. "Odor tracking in sharks is reduced under future ocean acidification conditions." *Global Change Biology* 21(4) (2015): 1454–1462, doi:10.1111/gcb.12678.

Emlen, Stephen T., and Lewis W. Oring. "Ecology, Sexual Selection and the Evolution of Mating Systems." *Science* 197(4300) (1977).

Engel, A., S. Thoms, U. Riebesell, E. Rochelle-Newall, and I. Zondervan. "Polysaccharide aggregation as a potential sink of marine dissolved organic carbon." *Nature* 428(6986) (2004): 929–932, doi:10.1038/nature02453.

Erisman, Brad E., Matthew T. Craig, and Philip A. Hastings. "A phylogenetic test of the size-advantage model: Evolutionary change in mating behavior influenced the loss of sex change in a fish lineage." *American Naturalist* 174 (2009): E83–E99.

Fadlallah, Yusef H. "Sexual reproduction, development and larval biology in scleractinian corals." *Coral Reefs* 2 (1983): 129–150.

Fairbairn, Daphne J. *Odd Couples: Extraordinary Differences between the Sexes in the Animal Kingdom.* Princeton, NJ: Princeton University Press, 2013.

Feldheim, K. A., D. D. Chapman, D. Sweet, S. Fitzpatrick, P. A. Prodohl, M. S. Shivji, and B. Snowden. "Shark virgin birth produces multiple, viable offspring." *J Hered* 101(3) (2010): 374–377, doi:10.1093/jhered/esp129.

Fitzpatrick, J. L., R. M. Kempster, T. S. Daly-Engel, S. P. Collin, and J. P. Evans. "Assessing the potential for post-copulatory sexual selection in elasmobranchs." *Journal of Biology* 80 (2012): 1141–1158.

Fong, Peter P., and Alex T. Ford. "The biological effects of antidepressants on the mollusks and crustaceans: A review." *Aquatic Toxicology* 151 (2014): 4–13.

Ford, A. T. "Intersexuality in Crustacea: An environmental issue?" *Aquat Toxicol* 108 (2012): 125–129, doi:10.1016/j.aquatox.2011.08.016.

Ford, A. T. "From gender benders to brain benders (and beyond!)." *Aquat Toxicol* 151 (2014): 1–3, doi:10.1016/j.aquatox.2014.02.005.

Ford, Alex T. "Can you feminize a crustacean?" *Aquatic Toxicology* 88 (2008): 316–321.

Forsyth, A. *A Natural History of Sex.* Buffalo, NY: Firefly Books, 2001.

Francisco, Rocha, Ángel Guerra, and Ángel F. González. "A review of reproductive strategies in cephalopods." *Biol Rev.* 76 (2001): 291–304.

Francisco, Rocha, Angel Gonzalez, Michel Segonzac, and Angel Guerra. "Behavioural observations of the cephalopod *Vulcanoctopus hydrothermalis.*" *CBM-Cahiers de Biologie Marine* 43(3–4) (2002): 299–302.

Grump, R. G., and R. H. Emson. "The nature history, life history and ecology of the two British species of *Asterina.*" *Field Studies* 5 (1983): 867–882.

Guest, James. "How reefs respond to mass coral spawning." *Science* 320(5876) (2008): 621-623.

Gusmao, L. F. M., and A. D. McKinnon. "Sex ratios, intersexuality and sex change in copepods." *Journal of Plankton Research* 31(9) (2009): 1101–1117, doi:10.1093/plankt/fbp059.

Haley, Leslie. "Sex determination in the American oyster." *Journal of Heredity* 68(2) (1977): 114–116.

Hall, V. R., and T. P. Hughes. "Reproductive strategies of modular organisms: Comparative studies of reef-building corals." *Ecology* 77(3) (1996): 950–963.

Hanlon, Roger T. "Mating systems and sexual selection in the squid *Loligo:* How might commercial fishing on spawning squids affect them?" *CalCOFI Report* 39 (1998).

Hanlon, Roger T., Malcolm J. Smale, and Warwick H. H. Sauer. "The mating system of the squid *Loligo vulgaris Reynaudii* (Cephalopoda, Mollusca) off South Africa: Fighting, guarding, sneaking, mating and egg laying behavior." *Bulletin of Marine Science* 71(1) (2002): 331–345.

Harrison, Peter L. "Sexual reproduction of scleractinian corals." In *Coral Reefs: An Ecosystem in Transition.* Ed. Z. Dubinsky and N. Stambler. New York, Springer, 2011, 59-86, doi:10.1007/978-94-007-0114-4_6.

Hastings, P. A., and C. W. Petersen. "Parental Care, Oviposition Sites, and Mating Systems in Blennioids." In *Reproduction and Sexuality in Marine Fishes: Patterns and Processes,* edited by K. S. Cole, 91–116. Oakland: University of California Press, 2010.

Hawkins, J. "Effects of fishing on sex-changing Caribbean parrotfishes." *Biological Conservation* 115(2) (2004): 213–226, doi:10.1016/s0006-3207(03)00119-8.

Hays, Graeme C., Jonathan D. R. Houghton, and Andrew E. Myers. "Pan-Atlantic leatherback turtle movements." *Nature,* June 3, 2004.

Hays, Graeme C., Jonathan D. R. Houghton, Andrew E. Myers. "Endangered species: Pan-Atlantic leatherback turtle movements." *Nature,* June 3, 2004.

Heyman, W. D., L. M. Carr, and P. S. Lobel. "Diver ecotourism and disturbance to reef fish spawning aggregations: It is better to be disturbed than to be dead." *Marine Ecology Progress Series* 419 (2010): 201–210, doi:10.3354/meps08831.

Heyman, William D., and Björn Kjerfve. "Characterizations of transient multi-species reef fish spawning aggregation at Gladden Spit, Belize." *Bulletin of Marine Science* 83(3) (2008): 531–551.

Hinck, J. E., V. S. Blazer, C. J. Schmitt, D. M. Papoulias, and D. E. Tilitt. "Widespread occurrence of intersex bass found in U.S. rivers." *Aquatic Toxicology* 94(4) (2009).

Hixon, Mark, David Conover, and Leesa Cobb. "Big old fat fecund female fish: The BOFFFF hypothesis and what it means for MPAs and fisheries management." *MPA News* 9(3) (2007).

Hoegh-Guldberg, Ove. "Climate change, coral bleaching and the future of the world's coral reefs." *Marine Freshwater Research* 50 (1999): 839–866, doi:10.1071/MF99078.

Hoelzel, A. Rus, Burney J. Le Boeuf, and Joanne Reiter Claudio Campagna. "Alpha-male paternity in elephant seals." *Behav Ecol Sociobiol* 46 (1999): 298–306.

Hoving, H. J., M. R. Lipinski, J. J. Videler, and K. S. Bolstad. "Sperm storage and mating in the deep-sea squid *Taningia danae* Joubin, 1931 (Oegopsida: Octopoteuthidae)." *Mar Biol* 157(2) (2010): 393–400, doi:10.1007/s00227-009-1326-7.

Hoving, Hendrik Jan T., Marek R. Lipiński, and Lammertjan Dam. "The male reproductive strategy of a deep-sea squid: Sperm allocation, continuous production, and long-term storage of spermatophores in *Histioteuthis miranda.*" *ICES Journal of Marine Science* 67 (2010): 1478–1486.

Hull, M. Q., A. W. Pike, A. J. Mordue (Luntz), and G. H. Rae. "Patterns of pair formation and mating in an ectoparasitic caligid copepod *Lepeophtheirus salmonis* (Krüyer 1837): Implications for its sensory and mating biology." *Royal Society* 353 (1998): 753–764.

Hunter, J., and J. Banks. "Observations on the structure and oeconomy of whales. By John Hunter, Esq. F. R. S.; communicated by Sir Joseph Banks, Bart. P. R. S." *Philosophical Transactions of the Royal Society of London* 77(0) (1787): 371–450, doi:10.1098/rstl.1787.0038.

Hylland, Ketil, and A. Dick Vethaak. 2011. "Impact of Contaminants on Pelagic Ecosystems." In Dam et al., *Ecological Impacts of Toxic Chemicals,* 212–223.

International Fund for Animal Welfare. *Breaking the silence: How our noise pollution is harming whales.* Sydney: IFAW Australia, 2013.

Ishimatsu, Atsushi, Masahiro Hayashi, Kyoung-Seon Lee, Takashi Kikkawa, and Jun Kita. "Physiological effects on fishes in a high-CO2 world." *Journal of Geophysical Research* 110(C9) (2005), doi:10.1029/2004jc002564.

Jarne, Philippe, and Josh R. Auld. "Animals mix it up too: The distribution of self-fertilization among hermaphroditic animals." *Evolution* 60(9) (2006): 1816–1824.

Jersabek, C. D., M. S. Luger, R. Schabetsberger, S. Grill, and J. R. Strickler. "Hang on or run? Copepod mating versus predation risk in contrasting environments." *Oecologia* 153(3) (2007): 761–773, doi:10.1007/s00442-007-0768-1.

Kawaguchi, S., R. Kilpatrick, L. Roberts, R. A. King, and S. Nicol. "Ocean-bottom krill sex." *J Plankton Res* 33(7) (2011): 1134–1138, doi:10.1093/plankt/fbr006.

Kelly, D. A. "Penises as variable-volume hydrostatic skeletons." *Ann N Y Acad Sci* 1101 (2007): 453–463, doi:10.1196/annals.1389.014.

Kelly, Lisa S., Terry W. Snell, and Darcy J. Lonsdale. "Chemical communication during mating of the harpacticoid *Tigriopus japonicus.*" *Royal Society* 353 (1998): 737–744.

Kennedy, Victor S., and Linda L. Breisch. *Maryland's Oysters Research and Management.* College Park: Maryland Sea Grant College, University of Maryland, 2001.

Klimley, Peter A. "Shark trails of the eastern Pacific." *American Scientist* 103 (June/August 2015).

Klimley, Peter A., *The Secret Life of Sharks: A Leading Marine Biologist Reveals the Mysteries of Shark Behavior.* New York: Simon & Schuster, 2003.

Knowlton, N. "The future of coral reefs." *Proc Natl Acad Sci USA* 98(10) (2001): 5419–5425, doi:10.1073/pnas.091092998.

Lange, Rolanda, Johanna Werminghausen, and Nils Anthes. "Does traumatic secretion transfer manipulate mating roles or reproductive output in a hermaphroditic sea slug?" *Behavioral Ecology and Sociobiology* 67(8) (2013): 1239–1247, doi:10.1007/s00265-013-1551-4.

Larson, Shawn, Jeff Christiansen, Denise Griffing, Jimiane Ashe, Dayv Lowry, and Kelly Andrews. "Relatedness and polyandry of sixgill sharks, *Hexanchus griseus,* in an urban estuary." *Conservation Genetics* 12(3) (2010): 679–690, doi:10.1007/s10592-010-0174-9.

Leenhardt, Pierre, Bertrand Cazalet, Bernard Salvat, Joachim Claudet, and François Feral. "The rise of large-scale marine protected areas: Conservation or geopolitics?" *Ocean & Coastal Management* 85 (2013): 112–118, doi:10.1016/j.ocecoaman.2013.08.013

Lehtonen, T. K., and K. Lindstrom. "Females decide whether size matters: Plastic mate preferences tuned to the intensity of male-male competition." *Behavioral Ecology* 20(1) (2008): 195–199, doi:10.1093/beheco/arn134.

Levitan, Don R. "Density-Dependent selection on gamete traits in three congeneric sea urchins." *Ecology* 83(2) (2002): 464–479, doi:10.2307/2680028.

Levitan, Don R., Hiroonobu Fukami, Javier Jara, David Kline, Tamara M. McGovern, Katie E. McGhee, Cheryl A. Swanson, and Nancy Knowlton. "Mechanisms of reproductive isolation among sympatric broadcast-spawning corals of the *Montastraea annularis* species complex." *Evolution* 58(2) (2004): 308–323.

Lobel, Phillip S., and Steve Neudecker. "Diurnal periodicity of spawning activity by the Hamlet Fish, *Hypoplectrus guttavarius* (Serranidae)." In *The Ecology of Coral Reefs* (NOAA Symposium Series Undersea Research, vol. 3). Ed. M. L. Reake. Rockville, MD: NOAA Undersea Research Program, 71–86.

Longhurst, Alan. "Murphy's law revisited: Longevity as a factor in recruitment to fish populations." *Fisheries Research* 56 (2002): 125–131.

Loosanoff, Victor L. "Seasonal gonadal changes in the adult oysters, *Ostrea virginica,* of Long Island Sound." *Biological Bulletin* 82(2) (1942): 195–206.

Lorio, Wendell J., and Sandra Malone. *The Cultivation of American Oysters* (Crassostrea virginica). Stoneville, MS: Southern Regional Aquaculture Center, 1994.

Luckherst, Brian E. "Evaluation of fisheries management and conservation measures taken to protect grouper spawning aggregations in the wider Caribbean: Case studies of Bermuda, Belize and Cayman Islands." *Proceedings of the Gulf and Caribbean Fisheries Institute* 58 (2007): 281–282.

Marshall, Dustin J., Selina S. Heppell, Stephan B Munch, and Robert R. Warner. "The relationship between maternal phenotype and offspring quality: Do older mothers really produce the best offspring?" *Ecology* 91(10) (2010): 2862–2873.

McClenachan, L., B. P. Neal, D. Al-Abdulrazzak, T. Witkin, K. Fisher, and J. N. Kittinger. "Do community supported fisheries (CSFs) improve sustainability?" *Fisheries Research* 157 (2014): 62–69.

Meek, Alexander. "The reproductive organs of Cetacea." *J Anat.* 52(2) (1918): 186–210.

Millus, Susan. "Dead, live guppies vie for paternity." *Science News* July 13, 2013.

Milner, J. M., E. B. Nilsen, and H. P. Andreassen. "Demographic side effects of selective hunting in ungulates and carnivores." *Conserv Biol* 21(1) (2007): 36–47, doi:10.1111/j.1523-1739.2006.00591.x.

Mohanty, Sobhi, Alfredo F. Ojanguren, and Lee A. Fuiman. "Aggressive male mating behaviors depends on female maturity in *Octopus bimaculoides.*" *Marine Biology* 161 (2014): 1521–1530.

Montaigne, Fen. "Still waters: The global fish crisis." *National Geographic.* April 2007.

Morris, Robert W. "Clasping mechanism of the cottid fish *Oligocottus snyderi* Greeley." *Pacific Science* 10(3) (1956): 314–317.

Morton, Brain. "The biology and functional morphology of *Chlamydoconcha orcutti* with discussion on the taxonomic status of the Chlamydoconchacea (Mollusca: Bivalvia)." *J. Zool., Lond* 195 (1981): 81–121.

Moyer, Jack T., and Roger C. Steene. "Nesting behavior of the anemonefish *Amphirion polymnus.*" *Japanese Journal of Ichthyology* 26(2) (1979).

Munday, P. L., P. M. Buston, and R. R. Warner. "Diversity and flexibility of sex-change strategies in animals." *Trends Ecol Evol* 21(2) (2006): 89–95, doi:10.1016/j.tree.2005.10.020.

Munday, Philip. 2002. "Bi-directional sex change: Testing the growth-rate advantage model." *Behavioral Ecology and Sociobiology* 52(3) (2002): 247–254, doi:10.1007/s00265-002-0517-8.

Munk, Ole. "Histology of the fusion area between the parasitic male and the female in the deep-sea anglerfish *Neoceratias spinifer* Pappenheim, 1914 (Teleostei, Ceratioidei)." *Acta Zoologica* 81 (2000): 315–324.

National Oceanic and Atmospheric Administration Marine Debris Program. *Report on the occurrence and health effects of anthropogenic debris ingested by marine organisms.* Silver Spring, MD, 2014.

Norman, Mark D., Julian Finn, and Tom Tregenza. "Female impersonation as an alternative reproductive strategy in giant cuttlefish." *Royal Society* 266 (1999): 1347–1349.

Nosel, A. P., A. Caillat, E. K. Kisfaludy, M. A. Royer, and N. C. Wegner. "Aggregation behavior and seasonal philopatry in male and female sharks *Triakis semifasciata* along the open coast of Southern California, USA." *Marine Ecology Progress Series* 499 (2014): 157–175.

Ospina-Alvarez, N., and F. Piferrer. "Temperature-dependent sex determination in fish revisited: Prevalence, a single sex ratio response pattern, and possible effects of climate change." *PLoS One* 3(7) (2008): e2837, doi:10.1371/journal.pone.0002837.

Palmer, A. Richard. "Caught right-handed." *Nature,* December 7, 2006, 689–692.

Patterson, S. "Size and growth modification in clownfish." *Nature,* July 10, 2003.

Payne, P. Michael, David N. Wiley, Sharon B. Young, Sharon Pittman, Phillip J. Clapham, and Jack W. Jossi. "Recent fluctuations in the abundance of baleen whales in the southern Gulf of Maine in relation to changes in selected prey." *Fishery Bulletin* 88(4) (1990): 687–696.

Pennington, J. Timothy. "The ecology of fertilization of echinoid eggs: The consequences of sperm dilution, adult aggregation, and synchronous spawning." *Biological Bulletin* 169(2) (1985): 417–430.

Pew Charitable Trusts, The. "Conserving Atlantic Bluefin Tuna with Spawning Sanctuaries." *Ocean Science Brief.* October 26, 2010. http://www.pewtrusts.org/en/research-and-analysis /reports/2010/10/26/ocean-science-series-science-brief-conserving-atlantic-bluefin-tuna -with-spawning-sanctuaries.

Policansky, David. "Sex change in plants and animals." *Ann. Rev. Ecol. Syst.* 13 (1982): 471–495.

Popper, A. N., and M. C. Hastings. "The effects of human-generated sound on fish." *Integr Zool* 4(1) (2009): 43–52, doi:10.1111/j.1749–4877.2008.00134.x.

Portnoy, D. S., and E. J. Heist. "Molecular markers: Progress and prospects for understanding reproductive ecology in elasmobranchs." *Journal of Biology* 80 (2012): 1120–1140.

Powell, Dawn K., Paul A. Tyler, and Lloyd S. Peck. "Effect of sperm concentration and sperm ageing on fertilisation success in the Antarctic soft-shelled clam *Laternula elliptica* and the Antarctic limpet *Nacella concinna.*" *Marine Ecology-Progress Series* 215 (2001): 191–200.

Prager, Ellen. *Sex, Drugs and Sea Slime: The Oceans' Oddest Creatures and Why They Matter.* Chicago: University of Chicago Press, 2011.

Pratt, H. L., and J. C. Carrier. "A review of elasmobranch reproductive behavior with a case study on the nurse shark, *Ginglymostoma cirratum.*" *Environmental Biology of Fishes* 60 (2001): 157–188.

Prevedelli, D., and R. Zunarelli Vandini. "Survival, fecundity and sex ratio of *Dinophilus gyrociliatus* (Polychaeta: Dinophilidae) under different dietary conditions." *Marine Biology* 133 (1999): 231–236.

Quinn, James F., Stephen R. Wing, and Louis W. Botsford. "Harvest refugia in marine invertebrate fisheries: Models and applications to the red sea urchin, *Strongylocentrotus franciscanus*." *American Zoologist* 33(6) (1993): 537–550.

Randall, John E. "Contributions to the biology of the queen conch, *Strombus gigas*." *Bulletin of Marine Science* 14(2) (1964): 246–295.

Rattanayuvakorn, Sukjai, Pisut Mungkornkarn, Amara Thongpan, and Kannika Chatchavalvanich. "Gonadal development and sex inversion in saddleback anemonefish Amphiprion polymnus Linnaeus (1758)." *Nat. Sci.* 40 (2006): 196–203.

Reinhardt, K., N. Anthes, and R. Lange. "Copulatory wounding and traumatic insemination." *Cold Spring Harb Perspect Biol* 7(5) (2015), doi:10.1101/cshperspect.a017582.

Richardson, Darren L., Peter L. Harrison, and Vicki J. Harriott. "Timing of spawning and fecundity of a tropical and subtropical anemonefish (Pomacentridae: *Amphiprion*) on a high latitude reef on the east coast of Australia." *Marine Ecology-Progress Series* 156 (1997): 175–181.

Richmond, Robert H. "Reproduction and recruitment in corals: Critical links in the persistence of reefs." In *Life and Death of Coral Reefs*. Ed. C. E. Birkeland. New York: Chapman & Hall, 1997, 175–197.

Rideout, Rich M., and George A. Rose. "Suppression of reproduction in Atlantic cod *Gadus morhua*." *Marine Ecology Progress Series* 320 (2006): 267–277.

Robertson, D. Ross, and Robert R. Warner. "Sexual patterns in the labroid fishes of the western Caribbean, II: The parrotfishes (Scaridae)." *Smithsonian Contributions to Zoology* 255 (1978).

Rogers-Bennett, Laura, Peter L. Haaker, Tonya O. Huff, And Paul K. Dayton. "Estimating Baseline Abundances of Abalone in California for Restoration." In *Abalone Baselines In California:* CalCOFI Report 43 (2002): 97–111.

Ross, Robert M. "Reproductive behavior of the anemonefish *Amphiprion melanopus* on Guam." *Copeia* 1978 (1): 103–107.

Ross, Robert M., Thomas F. Hourigan, Marvin M. F. Lutnesky, and Ishwar Singh. "Multiple simultaneous sex changes in social groups of a coral-reef fish." *Copeia* 1990 (2): 427–433.

Rouse, Greg W., Nerida G. Wilson, Shana K. Goffredi, Shannon B. Johnson, Tracey Smart, Chad Widmer, Craig M. Young, and Robert C. Vrijenhoek. "Spawning and development in *Osedax* boneworms (Siboglinidae, Annelida)." *Marine Biology* 156(3) (2008): 395–405, doi:10.1007/s00227-008-1091-z.

Rouse, Greg, Nerida G. Wilson, Katrine Worsaae, and Rober C. Vrijenhoek. "A dwarf male reversal in bone-eating worms." *Current Biology* 25(2) (2015): 236–241.

Ruxton, Graeme D., and David M. Bailey. "Combining motility and bioluminescent signalling aid, mate finding in deep sea fish: A simulation study." *Marine Ecology-Progress Series* 293 (2005): 253–262, doi:10.3354/meps293253.

Sadovy de Mitcheson, Yvonne, and Min Liu. "Functional hermaphroditism in teleosts." *Fish and Fisheries* 9 (2008): 1–43.

Sadovy, Y., M. Kulbicki, P. Labrosse, Y. Letourneur, P. Lokani and T. J. Donaldson. "The humphead wrasse, *Cheilinus undulatus:* Synopsis of a threatened and poorly known giant coral reef fish." *Reviews in Fish Biology and Fisheries* 13 (2003): 327–364.

Sadovy, Yvonne, and Douglas Y. Shapiro. "Criteria for the diagnosis of hermaphroditism in fishes." *Copeia* 1987 (1): 136–156.

Sakai, Yoichi, Masanori Kohda, and Tetsuo Kuwamura. "Effect of changing harem on timing of sex change in female cleaner fish *Labroides dimidiatus*." *Animal Behaviour* 62(2) (2001): 251–257, doi:10.1006/anbe.2001.1761.

Sala, Enric, Enric Ballesteros, and Richard M. Starr. "Rapid decline of Nassau grouper spawning aggregations in Belize: Fishery management and conservation needs." *Fisheries* 26(10) (2001): 23–30.

Santos, R. S. "Allopaternal care in the redlip blenny." *Journal of Fish Biology* 47 (1995): 350–353.

Searcy, William A. "The evolution effects of male selection." *Ann Rev. Ecol. Syst.* 13 (1982): 57–85.

Sella, Gabriella, and Liliana Ramella. "Sexual conflict and mating systems in the dorvilleid genus *Ophryotrocha* and the dinophilid genus *Dinophilus*." *Hydrobiologia* 402 (1999): 203–213.

Siveter, David J., G. Tanaka, C. Ú. Farrell, M. J. Martin, Derek J. Siveter, and D. E. G. Briggs. "Exceptionally preserved ostracods with developmental brood care from the Ordovician." *Current Biology* 24(7) (2014): 801–806.

Slabbekoorn, H., N. Bouton, I. van Opzeeland, A. Coers, C. ten Cate, and A. N. Popper. "A noisy spring: the impact of globally rising underwater sound levels on fish." *Trends Ecol Evol* 25(7) (2010): 419–427, doi:10.1016/j.tree.2010.04.005.

Smith, Homer W. *From Fish to Philosopher.* Boston: Little, Brown, 1953.

Smolker, Rachel A., Andrew F. Richards, Richard C. Connor, and John W. Pepper. "Sex differences in patterns of association among Indian Ocean bottlenose dolphins." *Behaviour* 123(1–2) (1992): 38–69.

Snook, R. R. "Sperm in competition: Not playing by the numbers." *Trends Ecol Evol* 20(1) (2005): 46–53, doi:10.1016/j.tree.2004.10.011.

Starr, Michael, John H. Himmelman, and Jean-Claude Therrianult. "Direct coupling of marine invertebrate spawning with phytoplankton blooms." *Science* 247 (1990): 1071–1074.

Starr, R. M., E. Sala, E. Ballesteros, and M. Zabala. "Spatial dynamics of the Nassau grouper *Epinephelus striatus* in a Caribbean atoll." *Marine Ecology Progress Series* 343 (2007): 239–249, doi:10.3354/meps06897.

Stephens, Philip A., and William J. Sutherland. "Consequences of the Allee effect for behaviour, ecology and conservation." *Tree* 14(10) (1999): 401-405.

Stockley, P., M. J. G. Gage, G. A. Parker, and A. P. Møller. "Sperm competition in fishes: The evolution of testis size and ejaculate characteristics." *American Naturalist* 149(5) (1997): 933–954.

Strathmann, Richard R. "Why life histories evolve differently in the sea." *American Zoologist* 30(1) (1990): 197–207.

Taylor, Martin. *Population Viability Analysis for the Southern Resident Population of the Killer Whale* (Orcinus orca). Tucson, AZ: Center for Biological Diversity, 2001.

Tegner, Mia J., and Paul K. Dayton. "Sea urchin recruitment patterns and implications of commercial fishing." *Science* 196(4287) (1977): 324–3267.

Titelman, J., O. Varpe, S. Eliassen, and O. Fiksen. "Copepod mating: Chance or choice?" *Journal of Plankton Research* 29(12) (2007): 1023–1030, doi:10.1093/plankt/fbm076.

Todd, Christopher D., Rebecca J. Stevenson, Helena Reinardy, and Michael G. Ritchie. "Polyandry in the ectoparasitic copepod *Lepeophtheirus salmonis* despite complex precopulatory and postcopulatory mate-guarding." *Marine Ecology-Progress Series* 303 (2005): 225–234.

Toonen, R. J., T. Wilhelm, S. M. Maxwell, D. Wagner, B. W. Bowen, C. R. Sheppard, S. M. Taei, T. Teroroko, R. Moffitt, C. F. Gaymer, L. Morgan, N. Lewis, A. L. Sheppard, J. Parks, A. M. Friedlander, and Tank Big Ocean Think. "One size does not fit all: The emerging frontier in large-scale marine conservation." *Mar Pollut Bull* 77(1–2) (2013): 7–10, doi:10.1016/j.marpolbul.2013.10.039.

van Overzee, Harriët M. J., and Adriaan D. Rijnsdorp. "Effects of fishing during the spawning period: Implications for sustainable management." *Reviews in Fish Biology and Fisheries* 25(1) (2015): 65–83, doi:10.1007/s11160-014-9370-x.

Vasconcelos, R. O., M. C. Amorim, and F. Ladich. "Effects of ship noise on the detectability of communication signals in the Lusitanian toadfish." *J Exp Biol* 210(pt. 12) (2007): 2104–12, doi:10.1242/jeb.004317.

Velando, A., J. Eiroa, and J. Dominguez. "Brainless but not clueless: Earthworms boost their ejaculates when they detect fecund non-virgin partners." *Proc Biol Sci* 275(1638) (2008): 1067–72, doi:10.1098/rspb.2007.1718.

Vrijenhoek, R. C., S. B. Johnson, and G. W. Rouse. "Bone-eating *Osedax* females and their 'harems' of dwarf males are recruited from a common larval pool." *Mol Ecol* 17(20) (2008): 4535–44, doi:10.1111/j.1365-294X.2008.03937.x.

Warner, Robert R. "Mating behavior and hermaphroditism in coral reef fishes." *American Scientist* 72 (1984): 2 128–136.

Warner, Robert R., and Ronald K. Harlan. "Sperm competition and sperm storage as determinants of sexual dimorphism in the dwarf surfperch, *Micrometrus minimus*." *Evolution* 36(1) (1982): 44–55.

Warner, Robert R., and Steven G. Hoffman. "Population density and the economics of territorial defense in a coral reef fish." *Ecology* 61(4) (1980): 772–780.

Wedell, Nina, Matthew J. G. Gage, and Geoffrey A. Parker. "Sperm competition, male prudence and sperm-limited females." *Trends in Ecology & Evolution* 17(7) (2002): 313–320.

Whitney, Nicholas M., Harold L. Pratt, Jr., Theo C. Pratt, and Jeffery C. Carrier. "Identifying shark mating behaviour using three-dimensional acceleration loggers." *Endangered Species Research* (2010), doi:10.3354/esr00247.

Wiig, Oystein, Andrew E. Derocher, Matthew M. Cronin, and Janneche U. Skaare. "Female pseudohermaphrodite polar bears at Svalbard." *Journal of Wildlife Diseases* 34(4) (1998): 792–796.

Wolfe, Kennedy, Abigail M. Smith, Patrick Trimby, And Maria Byrne. "Vulnerability of the paper nautilus (*Argonauta nodosa*) shell to a climate-change ocean: Potential for extinction by dissolution." *Biol. Bull.* 223(2) (2012): 236–244.

Worsaae, K., and G. W. Rouse. "The simplicity of males: Dwarf males of four species of *Osedax* (Siboglinidae; Annelida) investigated by confocal laser scanning microscopy." *J Morphol* 271(2) (2010): 127–142, doi:10.1002/jmor.10786.

Wyatt, T. D., J. D. Hardege, and J. Terschak. "Ocean acidification foils chemical signals." *Science* 346(6206) (October 2014).

Wyatt, Tristram D. "Introduction to Chemical Signaling in Vertebrates and Invertebrates." In *Neurobiology of Chemical Communication.* Ed. C. Mucignat-Caretta. Boca Raton, FL: CRC Press, 2014, 1-22.

Wyatt, Tristram D. "How animals communicate via pheromones." *American Scientist* 103 (2015).

Xu, E. G., S. Liu, G. G. Ying, G. J. Zheng, J. H. Lee, and K. M. Leung. "The occurrence and ecological risks of endocrine disrupting chemicals in sewage effluents from three different sewage treatment plants, and in natural seawater from a marine reserve of Hong Kong." *Mar Pollut Bull* 85(2) (2014): 352–362, doi:10.1016/j.marpolbul.2014.02.029.

Yang, G., P. Kille, and A. T. Ford. "Infertility in a marine crustacean: Have we been ignoring pollution impacts on male invertebrates?" *Aquat Toxicol* 88(1) (2008): 81–7, doi:10.1016/j.aquatox.2008.03.008.

Yoshizawa, K., R. L. Ferreira, Y. Kamimura, and C. Lienhard. "Female penis, male vagina, and their correlated evolution in a cave insect." *Curr Biol* 24(9) (2014): 1006–1010, doi:10.1016/j.cub.2014.03.022.

Zeh Jeanne A., and David W. Zeh. "Reproductive mode and genetic benefits of polyandry." *Animal Behavior* 61 (2001): 1051–1063.

Zervomanolakis, I., H. W. Ott, D. Hadziomerovic, V. Mattle, B. E. Seeber, I. Virgolini, D. Heute, S. Kissler, G. Leyendecker, and L. Wildt. "Physiology of upward transport in the human female genital tract." *Ann N Y Acad Sci* 1101 (2007): 1–20, doi:10.1196/annals.1389.032.

Zhang, Dong, Qiang Lin, and Junda Lin. "Sex-dependent energetic cost of a protandric simultaneous hermaphroditic shrimp *Lysmata wurdemanni* under different social conditions." *Journal of Experimental Marine Biology and Ecology* 394 (2010): 134–140.

Zimmer, Carl. *Parasite Rex: Inside the Bizarre World of Nature's Most Dangerous Creatures.* New York: Free Press, 2000.

RESOURCES: DIVING A LITTLE DEEPER

If you're feeling inspired to learn more, or do more to promote the sex lives of sea life, I encourage you to check out the Sex in the Sea website (sexinthesea.org), where a wealth of up-to-date information awaits. For those looking for a cursory exploration to help satisfy that urge right now, a few starting resources are provided here. In no particular order, and just the tip of the iceberg, this list can help kindle a productive love affair with the ocean.

SAFE-FOR-SEX SEAFOOD

FishWatch
NOAA Fisheries site with accessible, science-based facts to help US consumers make smart, sustainable seafood choices.
fishwatch.gov

LocalCatch.org
A network of community-supported fisheries and small-scale harvesters with resources for consumers to find seafood nearby.
localcatch.org

The Safina Center
A nonprofit focused on creative communication of conservation issues, The Safina Center offers a sustainable seafood guide that includes information regarding mercury loads in fish.
safinacenter.org/seafoods/

Seafood Watch
Monterey Bay Aquarium's recommendations for selecting seafood fished or farmed in environmentally responsible ways.
seafoodwatch.org

GREAT OCEAN PORN (PG-13),
AKA STUNNING MARINE PHOTOGRAPHY

Bryce Groark Photography
Award-winning creative director, producer, and photographer based in Hawaii.
brycegroark.com

Klaus Stiefel: Pacificklaus
Neurobiologist Klaus M. Stiefel's blog, featuring his underwater photography, marine biology insights, and resources.
pacificklaus.com

Octavio Aburto-Oropeza
Marine biologist and nature photographer with a focus on marine ecosystems in Mexican coastal waters.
octavioaburto.com

Raph's Wall
Honolulu marine biologist and diver Raphael Ritson-Williams's blog with his photography, research, and writing.
raphswall.com

Tim Calver Photography
World-traveling freelance photographer with an extensive underwater portfolio.
timcalver.com

EDUCATION

Discovery of Sound in the Sea
A University of Rhode Island online resource with audio clips of many marine species' mating calls.
dosits.org

Ocean Portal
Smithsonian Institution National Museum of Natural History's educational site that focuses on everything ocean including organisms, art, and researchers.
ocean.si.edu/category/ocean-portal

Ocean Views
National Geographic blog bringing diverse voices together to discuss threats facing our ocean and celebrate successes.
voices.nationalgeographic.com/blog/ocean-views

Right Whale Listening Network
The Cornell Lab of Ornithology's bioacoustics research program records right whale sounds from a network of buoys.
listenforwhales.org

Sharks4Kids
The nonprofit website of photographer Jillian Morris provides shark education for teachers and students.
sharks4kids.com

Tagging of Pelagic Predators
An international, multidisciplinary collaboration that allows users to interact with oceanographic and animal tracking data.
gtopp.org

TAKING INITIATIVE: A FEW WAYS TO ENGAGE
MENTIONED IN THE TEXT

Coral Restoration Foundation
Nonprofit conservation organization with volunteer opportunities for creating offshore coral nurseries and restoring critically endangered coral reefs.
coralrestoration.org

Global Fishing Watch
An interactive web tool designed to show all trackable fishing activity in the ocean worldwide.
globalfishingwatch.org

Grunion Greeters
Pepperdine University–led collaborative research project with hundreds of volunteers to study the spawning activity and habitat of grunion.
grunion.pepperdine.edu/ggproject.htm

International Coastal Cleanup
The Ocean Conservancy's worldwide volunteer effort to clean waterways and the ocean.
oceanconservancy.org/our-work/international-coastal-cleanup/

Sustainable Seafood Week National*
Weeklong event series in select US cities that celebrates legal, traceable, renewable, and trustworthy seafood.
sswnational.com

---

*Organization or program with which I am affiliated.

INDEX

Italicized page numbers refer to endnotes.

abalone, 164, 185, 192–4, 201, 217–18, 231
adelphophagy (embryo cannibalism), 147–8
Allee effects, 193
amphipods, 92–5, 228
anal glands, 52
anemonefish. *See* clownfish (anemonefish)
angelfish, 85
anglerfish, 154–5, 157, *252*
antennules, 13, 38, 40, 44, *243*
antiarchs, 106
aquarium trade, 37, 47–8, 56, 60, 227
Aquilino, Kristin, 217
argonaut, 99, 124–6
Atema, Jelle, 38, 40, 44–5
automatic identification system (AIS), 220–1

barnacles, 54, 115–18, 121
Berghe, Eric van den, 50
Birkhead, Tim, 143, *250*
blennies, 52–3
    redlip blennies, 53
    tube blennies, 54, 56–7, 62
Blue Halo Initiative (Waitt Institute), 224
blue whales, 7, 8, 24–8, 61, 115–16, 222, *241. See also* whales
bluefin tuna, 2, 177–180
Bodega Marine Laboratory (University of California, Davis), 194, 217
BOFFFFs (big, old, fat, fecund female fish), 76–7, 79, 83, 86, 209, 231, *262*
boundary layers of the ocean, 11–12

bowerbirds, 51

Cabo Pulmo National Marine Park, 224–5
cephalopods, 63, 126. *See also* cuttlefish; octopuses; squid
Chapman, Demian, 147–9, 159–61, 222
Chondrichthyes (cartilaginous fishes), 129. *See also* sharks
Chromodoris reticulata (tropical nudibranch), 127–8
Clapham, Phillip, 26–7, 112–13, 152
climate change, 22, 170, 203, 209, 227, 230–1
    and coastline erosion, 20
    and ocean acidification, 3, 45
    and ocean surface temperature, 12
clownfish (anemonefish), 69, 71, 77–8, 87–9, *246*
Coalition to Restore Coastal Louisiana, 230
cod, 10, 37, 48, 57–60, 129, 228, *244*
coloration, 135, 175–7
    and clownfish, 77
    and cuttlefish, 63–6
    and horseshoe crabs, 172
    and mandarinfish, 55–7, 60
    and Nassau grouper, 14, 175
    and parrotfish, 74–5
    and pufferfish, 51
    and squid, 119
comb jelly (ctenophores, Beroë), 158, 186
Community Supported Fisheries (CSFs), 214

conchs, 101–2, 116, 126, 164
Convention on International Trade in
    Endangered Species (CITES), 221
copepods, 2, 10–13, 132, 196, *239*
corals, 13, 51, 128, 184, 233–4
    boulder star coral (*O. franksi,* "franks"),
        198–200, 234
    as endangered, 202–4, 230
    farming, 226–7
    lobed star coral (*Orbicella annularis*),
        198–200
    and parrotfish, 74–5, 219
    and spawning, 1, 3, 194–204
    and zooxanthellae, 202–3
Cowan, Diane, 40
Cranford, Ted, 28, *242*
ctenophores (comb jellies, Beroë), 158,
    186
cuttlefish, 35, 63–5, 119

Darwin, Charles, 117, 135, 190
dating. *See* mating tactics and rituals
density, 164, 180, 191–4, 201
diatoms, 91–2
DNA, 148, 159, 161, 171, 190, 221–2
    microsatellites (short repetitive patterns
        of DNA), 23
dolphins, 26, 111–13, 137–9, 176, *249*
domoic acid poisoning, 91–2
Dutton, Peter, 21, *241*

eels, 75, 173
ejaculation, 68, 105, 122, 128, 130–1,
    155–7, 164, 186, 188. *See also*
    sperm
electromagnetism, 16–18, 189
elephant seals, 18–20, 72, 145–6, 153,
    241. See also seals
embryos, 46, 100, 145, 159, 168, 190
    embryo cannibalism (adelphophagy),
        148–50
endocrine disrupting chemicals (EDCs),
    94–5, 227–9
epitoky, 126–7
Erisman, Brad, 175–7, 180

Feldheim, Kevin, 23, 147, 149, 160, 222
feminization, 94, 227. *See also*
    masculinization

fertilization, 66–7, 79, 83
    of corals, 199–202
    and critical mass, 193
    of cuttlefish, 63–4
    external, 48–50, 52, 56, 162, 171
    and female control, 12, 136, 146,
        154–62
    fertilization trials (coral), 199–200
    and hermaphrodites, 77, 79, 83
    internal, 100–1, 106, 110, 114, 117,
        130, 132, 144, 147, 198, *248*
    of lobsters, 41, 44
    of sea urchins, 185–92
    of seahorses, 46
    self-fertilization, 103–4
    and sperm storage, 123, 144
    and timing, 142–3, 155
    of wrasse, 66–7
Field Museum of Natural History, 23, 222
fishing and harvesting
    3-D ocean farming, 216
    of abalone, 194
    almadrabas (large seine nets), 179, *253*
    alternatives to, 225–7
    aquarium trade, 37, 47–8, 56, 60, 227
    of Atlantic silversides, 87
    bottom trawling, 60, 221, *240*
    bycatch, 138–9, 180, 212
    of cetaceans, 113
    of cod, 60
    farming, 216–18
    of grunions, 168
    of haddock, 58, 60
    hook and line, 15, 60
    of horseshoe crabs, 173–4
    illegal, 174, 180, 213–15, 220–1
    and large-scale marine protected areas,
        218–20
    of Nassau grouper, 177, 223
    overfishing, 86, 180, 191, 203, 209–12,
        219, 224, 231, *247*
    responsible and sustainable, 56, 212–17,
        209–25, 236
    of sea turtles and eggs, 22
    of seahorses, 47–8
    and seamounts, 17–18
    selective, 60–1, 80, 86–7, 209–12, 247
    shrinkage 86, 87, 210
    of sharks, 147, 209–10, 221–2, 226

and slave labor, 213, 215, *255–6*
of smalltooth sawfish, 160, 161
and spawning aggregation, 15, 18, 177, 180, 210, 223
of toothed whales, *251*
transparency and traceability, 215
of tuna, 180, *253*
See also aquarium trade
flatworms, 99, 102–6, 124
flying fish, 212
Fogarty, Nicole, 199
Followfish, 214
Ford, Alex, 92–5, 228
Franks, Peter, 10, *239*
Future of Fish, 213

gametes, 15, 18, 60, 79, 105, 126, 160, 163, 169, 176, 186–7, 191, 193, 195–201, 203–4
Gelsleichter, Jim, 160–1
genital gender bias, 134–6
Global Fishing Watch, 221
gonopodia, 110
gonopods, 42
Great Barrier Reef Marine Park, 218–19
Grober, Matthew, 88, 90, *245*
Grouper Moon Project, 13, 15, 223, *240*
groupers, 69, 81, 219, 224
gag groupers, 225
Nassau groupers, 13–16, 174–5, 177, 223
red hind, 211
Grubbs, Dean, 130
Gulf pipefish, 150
Gulf Wild, 214–15
guppies, 110, 144, 145, 228

haddock, 57–60, *244*
Harding, Juliana, 79, 200–1
harvesting. *See* fishing and harvesting
Hastings, Phil, 52–4, 57, 62
hectocotyli, 119–20, 122, 124–6
Helmholtz Centre for Ocean Research, 121
hemipenes, 108
hermaphrodites, 71–6, 80, 103–5, 117, 196
bluebanded gobies, 88–90
clownfish, 69, 71, 77–8, 87–9, *246*

common slipper shell snails, 43–4, 73, 94
flatworms, 99, 102–6, 124
gag groupers, 225
oysters, 69, 78–80, 86–7, 91, 94
parrotfish, 73–5, 83–4
polychaete worm (*Ophryotrocha puerilis*), 87–8
and prey, parasites, and pollutants, 91–5
protandrous, 73, 75–80, 87
protogynous, 73, 75, 80–2, 83, 86, 87, 225
ribbon eels, 75
rusty angelfish, 85
sea goldies, 82–3
sequential, 43, 73–4, *245*
and sex-change reversals (flip-flops), 87–91
and shape shifting, 75
and size-selective fishing, 86–7
simultaneous, 43, 73
wrasse, 75, 81–2
horseshoe crabs, 165, 170–4, 186, *252*
Hoving, Henk-Jan, 121–3
hybrids, 198

illegal, unreported, and unregulated (IUU) fish, 214. *See also* fishing and harvesting: illegal
in vitro fertilization (IVF), 183, 187
International Coastal Cleanup (Ocean Conservancy), 229

Klimley, Peter, 17

leks, 58
Levitan, Don, 191–2
Lobster Conservancy, 40, *242*
lobsters, 37–45, *243*
antennules, 38, 40, 44, *243*
molting, 40–3
swimmerettes, 40, 42, *243*
love envelopes, 12

mandarinfish, 55–7, 60
marine protected areas (MPAs), 218–20
Marine Stewardship Council (MSC), 213
masculinization, 94, 99, 227–9. *See also* feminization

Mate, Bruce, 113
mating tactics and rituals, 5, 8–9m
    36–7
  and coloration, 14
  and dance, 55–7, 57–63
  and electromagnetism, 16–18
  and fatherhood, 45–50
  and nests, 50–4
  and place (homing), 18–24
  and scent, 37–45
  "singles bars," 7, 9–16, 196, *239*
  on the sly, 63–8
  and "sneakers," 64,68, 119–20
  and sound, 24–31, 57–63
  and spawning aggregations, 14–16
  and splash displays, 31–3
  and urine, 36, 37–40, 42, 44
  See also sexual intercourse
Mesnick, Sarah, 134, 136–41, *242, 250*
microsporidian, 93
mobula rays, 9, 31–3, 201
molting, 40–3
monogamy, 42, 45, 77, 166
Monterey Bay Aquarium, 213
mosquitofish, 110, 116
Mote Marine Lab, 227

Nassau grouper, 13–16, 174–5, 177, 223
National Oceanic and Atmospheric
    Administration (NOAA), 26, 112,
    134, 152, 229
nests, 22, 30–1, 37, 49–53, 62, 65, 89,
    170–1
Nets to Energy, 229
nudibranch, 127–8

ocean acidification, 3, 45. *See also* climate
    change
Ocean Conservancy, 229
ocean zoning, 223–5
octopuses, 63, 118–19, 124–5
  argonaut, 99, 124–6
ostracods, 107–10, 116–17, 132
  Luprisca incuba, 110
Oyster Recovery Partnership (ORP), 230
oysters, 69, 78–80, 86–7, 91, 94, 124, 184,
    188, 195
  oyster reefs, 3, 9, 78–9, 201, 204,
    229–30, *246*

Pacific Remote Islands Marine National
    Monument, 219–20
Palau, 220–1, 226
parasites, 32, 93–5, 126
parrotfish, 231
  bucktooth parrotfish, 83–4
  bumphead parrotfish, 219
  queen parrotfish, 73–5
parthenogenesis (type of asexual
    reproduction), 158–61
Patut, Wayan, 226–7
penis, 100–102
  baculum (internal penis bone), 111
  of barnacles, 115–18, 121
  claspers (shark penises), 128–33
  of conchs, 101–2
  and dismemberment, 124–8
  of flatworms, 99, 102–6, 124
  Colymbosathon ecplecticos (oldest known
      fossil penis), 106–7
  and flexibility, 110–14
  of human beings, 110–11
  and length, 114–18
  of mammals, 111–12
  of ostracods, 107–10, 116–17, 132
  penis fencing, 99, 102–6
  purpose of, 100–101
  of sharks, 128–32
  verge (conch penis), 101, 116
  of whales, 115–16
  See also sperm
pH levels, 44–5
pheromones, 11, 43–5, 93–4
phytoplankton, 187–8, 200
pipefish, 150
pollutants, 3, 30, 45, 94–5, 114, 203–4,
    209, 216, 227–8, 230
pregnancy, 23–4, 113, 133, 160–1, *242*
  and diapause, 145
  male, 46–8
  of sharks, 133
  and sperm storage, 142
  and stress, 204
Project Eyes on the Seas, 220–1
pufferfish, 111, 172
  whitespotted pufferfish, 51–2

rays, 23, 129–31, 147, 210
  Javanese cownose ray, 143–4

manta rays, 226
mobula rays, 9, 31–3, 201
smalltooth sawfish, 160–1
stingrays, 16, 131–2
reproduction. *See* sexual intercourse
Rouse, Greg, 155–6
Ruiz, Rob, 214

salmon, 2, 18, 23, 68, 129, 211, 216
sardines, 3, 9, 10, 210, 213
scent, 21, 36–45, 67–8
sea bass, 105, 212
sea cucumbers, 185–6
sea goldie (lyretail *anthias*), 82–3
sea hares, 104
sea slugs, 127–8
sea stars, 72, 126, 185–6
sea turtles, 20–4, 71, 93, 136, 142–3, 146, 169, 210, 233
sea urchins, 94, 164, 183–4, 185–8, 190–3, 201
seahorses, 45–8, 77, 208
seals, 111, 129, 169
    elephant seals, 18–20, 72, 145–6, 153, *241*
    ringed seals, 227
seamounts, 16–18, 210, 224–5, *240*
Seaver, Barton, 212–13
Scripps Institution of Oceanography, 10, 32, 52, 155
Seafood Watch (Monterey Bay Aquarium), 213
Semmens, Brice, 15, *240*
sex changers. *See* hermaphrodites
sex-sea soundtracks, 8, 36, 70, 100, 134, 166, 184
sex slaves, 134, 154–7
sexual intercourse, 10, 97, 100–1
    and ancient appendages, 106–8
    of cephalopods, 118–24
    of *Chromodoris reticulata* (sea slugs), 127–8
    and death, 20, 120–21, 153
    and female choice, 135–6, 145–6, 158–62
    and flexible penises, 110–14
    and injuries, 146, 153
    length of, 145
    and romance (lobsters), 42

unwanted, 141, 149
    See also fertilization; mating tactics and rituals; penis; sperm; testes; vagina
sexual success
    and density, 164, 180, 191–4, 201
    and timing, 194–204
shape shifting, 64, 75, 117
sharks, 2, 4, 31, 136, 160, 176
    and adelphophagy (embryo cannibalism), 147–8
    blacktip sharks, 159
    blue sharks, 130, 131
    bonnethead sharks, 149–50, 159
    brownbanded bamboo sharks, 143
    cat sharks, 147
    and claspers (penises), 128–33
    and electromagnetism, 16–17
    hammerhead sharks, 16–17, 20, 149–50, 221
    and homing, 23–4
    lemon sharks, 23–4, 147, 149
    and marine reserves, 218–19
    nurse sharks, 131–2
    and overfishing, 147, 209–10
    and parthenogenesis, 159
    sand tiger sharks, 131, 146–51
    shark fins, 221–2, 226
    and sperm selection, 147–51
    and sperm storage, 133, 143–4
    whitespotted bamboo sharks, 159
    zebra sharks, 159
shrimp, 70, 92, 197, 211, 213, 216–17, *247*
Siveter, David, 107–8, 110
SkyTruth, 221
Smith, Bren, 216
snails
    abalone, 164, 185, 192–4, 201, 217–18, 231
    common slipper shell snails, 43–4, 73, 94
    conch, 101–2, 116, 126, 164
    sea hares, 104
    and masculinization, 94, 99, 227–9
snapper, 3, 10, 129, 208, 209, 224
    spawning aggregations, 174–6
    vermillion snapper, 76
sound, 24–33, 57–8

spawning
  aggregations, 14–16, 18–19, 173–7,
    180–1, 210–11, 213, 223–5
  broadcast, 186, 190–1
  and coloration, 14
  of corals, 1, 3, 194–204
  and fishing and harvesting, 15, 18, 177,
    180–1, 210–11, 223–5
  of Nassau grouper, 14–16
  of salmon, 18
  of wrasse, 2
sperm
  competition, 151–3
  ejaculation, 68, 105, 122, 128, 130–1,
    155–7, 164, 186, 188
  of ostracod, 108
  and parthenogenesis (type of asexual
    reproduction), 158–61
  polyspermy (multiple sperm penetrating
    an egg), 158, 186, 190–2
  selection, 145–51
  storage, 120, 123, 130, 142–5
  of whales, 151–2
  See also fertilization; penis
spermatophore, 12–13, 42, 118–19,
  122–23
squid, 4, 65, 100, 118–24, 210, 218
Stewart, Josh, 32
stingrays, 16, 131–2
Storied Fish, 213–15, *256*
swim bladder, 57–8

terminal organs, 120–3
testes
  of hermaphrodites, 75
  and size, 114, 134, 140, 151, 147, *250–1*
  of squid, 65–6
  of whales, 114, 139–40, 151–2, 157
ThisFish, 214
tributylin (TBT), 94, 229
tuna, 211, 214
  bluefin tuna, 2, 178–80, 212

underwater ecotourism industry, 226–7
urine, 36, 37–40, 42, 100, *243*

"golden shower," 36, 42

vaginas, 114, 162, 163
  of blue sharks, 130
  and genital gender bias, 134–6
  of ostracods, 108
  of *Pisione* worms, 132
  of sea hares, 104
  of whales, 133, 136–41, *250*
Vecchione, Michael, 121, 123
virgin birth, 134, 158–9, 161
Vrijenhoek, Bob, 155

walrus, 111
Warner, Bob, 49–50, 62, 67, *244*
whales
  blue whales, 7, 8, 24–8, 61, 115–16,
    222, *241*
  gray whales, 25, 27, 113
  humpback whales, 25, 61–3, 116, 152,
    231
  and *Osedax,* 155–7, 252
  right whales, 25, 112–14, 151–2, 222
  sperm whales, 27–9, 113, 120, 141,
    242
  surface active groups (SAGs), 152
  testes of, 114, 139–40, 151–2, 157
  vaginas of, 133, 136–41, *250*
White, Barry, 7, 24, 217
worms, 134
  Dinophilus gyrociliatus, 157
  flatworms, 99, 102–6, 124
  Osedax, 155–7, 252
  Pisione worms, 132
  ragworm, 126–7
  segmented palolo worm, 126
  spoon worms, 157
wrasse, 2, 67, 197
  and alternative male types, 65–7
  bluehead wrasse, 65–7
  cleaner wrasse, 81–2
  Maori wrasse (humphead), 75
  peacock wrasse, 49–50, 62

Yen, Jeannette, 10, 12, *239*